Fundamental of Biochemical Engineering

Fundamental of Biochemical Engineering

Kurra Venkata Gopiah

RANDOM

RANDOM PUBLICATIONS

NEW DELHI (INDIA)

Fundamental of Biochemical Engineering

ISBN 978-93-5111-797-1

Published in 2016 in India by
RANDOM PUBLICATIONS

4376-A/4B, Gali Murari Lal, Ansari Road
New Delhi-110 002
Phone : +9111-43580356, 011-23289044, 011-43142548
e-mail: sales@randompublications.com,
info@randompublications.com, randomexports@gmail.com

Reprinted 2026

Type Setting by : Friends Media, Delhi-110089

Preface

Biochemical engineering/ biotechnology engineering/bioprocess engineering is a branch of chemical engineering that mainly deals with the design and construction of unit processes that involve biological organisms or molecules, such as bioreactors. Its applications are in the petrochemical industry, food, and pharmaceutical, biotechnology, and water treatment industries. Biochemical engineers translate exciting discoveries in life sciences into practical materials and processes contributing to human health and well-being. If you are interested in applying your skills and knowledge to meet global challenges relating to the development of novel medicines, pioneering stem cell therapies or green sustainable technologies, then Biochemical Engineering at UCL is for you. Biochemical engineers are key players in the greatest biomedical challenges that lie ahead including the preparation of vaccines for the treatment of diseases as diverse as HIV, AIDS, swine flu and cervical cancer as well as the construction of tissue engineered replacement body parts. Crucially, they are continually concerned with ensuring that the methods used to prepare the medicine are the safest possible, as even a few virus molecules from a contaminated source are enough to make a valuable medicine a dangerous vehicle for infection, especially if it is given by injection. In the next decade and beyond, the contributions of biochemical engineers to the worldwide community will be vital not only in the area of new medicines but also for novel materials from renewable sources such as biofuels, improved approaches to reducing and dealing with environmental pollution and more nutritious foods.

– Author

Contents

1

Process Biotechnology

Biotechnology is "the integration of natural sciences and engineering in order to achieve the application of organisms, cells, parts thereof and molecular analogues for products and services". Environmental biotechnology as discussed in this briefing paper is the application of these processes for the protection and restoration of the quality of our environment. This briefing paper reviews the various areas of environmental biotechnology together with their related issues and implications.

The overall aim is to provide balanced information and advance public debate. This paper results from the combined contributions of scientists, industrialists, and governmental and environmental organisations across Europe. It is intended to supply information and does not represent the views or policy of the European Federation of Biotechnology or any other body.

Biotechnological processes to protect the environment have been used for almost a century now, even longer than the term 'biotechnology' exists. Municipal sewage treatment plants and filters to purify town gas were developed around the turn of the century. They proved very effective although at the time, little was known about the biological principles underlying their function. Since that time our knowledge base has increased enormously. It also deals with the societal aspects of environmental biotechnology.

Biotechnological techniques to treat waste before or after it has been brought into the environment are described and exemplified in the section on bioremediation.

Biotechnology can also be used to develop products and processes that generate less waste and use less non-renewable resources and energy. In this respect biotechnology is well positioned to contribute to the development of a more sustainable society, a principle which was advocated in the Brundtland Report in 1987 and in Agenda 21 of the second Earth Summit in Rio de Janeiro in 1992 and which has been widely accepted in the mean time.

The section on prevention deals with this subject. Biotechnological techniques to monitor the quality of the environment are presented in the section on detection and monitoring. Recombinant DNA technology has improved the possibilities for the prevention of pollution and holds a promise for a further development of bioremediation.

The development of modern biotechnology has been accompanied by the establishment or adaptation of regulations to deal with genetically modified organisms. What this means for environmental biotechnology is embodied in the section on legislation. The section on public opinion, dialogue and debate highlights how people feel about environmental biotechnology and ways in which their opinion is influenced.

BIOREMEDIATION

Bioremediation is the use of biological systems for the reduction of pollution from air or from aquatic or terrestrial systems. Microorganisms and plants are the biological systems which are generally used. Biodegradation with micro-organisms is the most frequently occurring bioremediation option.

Micro-organisms can break down most compounds for their growth and/ or energy needs. These biodegradation processes may or may not need air. In some cases, metabolic pathways which organisms normally use for growth and energy supply may also be used to break down pollutant molecules.

In these cases, known as co-metabolism, the microorganism does not benefit directly. Researchers have taken advantage of this phenomenon and use it for bioremediation purposes. A complete biodegradation results in detoxification by mineralising pollutants to carbon dioxide, water and harmless inorganic salts.

Incomplete biodegradation will yield breakdown products which may or may not be less toxic than the original pollutant. Incomplete biodegradation of tri- or tetrachloroethylene for instance can yield vinylchloride, which is more toxic and carcinogenic than the original compounds.

Biodegradation may occur spontaneously, in which case the expressions "intrinsic bioremediation" or "natural attenuation" are often used. In many cases however the natural circumstances are not favourable enough for this to happen due to the lack of enough nutrients, oxygen or suitable bacteria. Such situations may be improved by supplementing one or more of these prerequisites. Extra nutrients were for instance disseminated to speed up the breakdown of the oil spilled on 1000 miles of Alaskan shoreline by the super tanker Exxon Valdez in 1989.

The future trend in bioremediation increasingly is to look at the speed of unaided biodegradation first and act only if there is insufficient activity to remove the pollutant quickly enough to prevent any expected risks of the pollutant. Bioremediation techniques can be used to reduce or to remove hazardous waste which has already polluted the environment.

They can also be used to treat waste streams before they leave production facilities: end-of-pipeprocesses.

Waste Water and Industrial Effluents

Micro-organisms in sewage treatment plants remove the more common pollutants from waste water before it is discharged into rivers or the sea. Increasing industrial and agricultural pollution has led to a greater need for processes that remove specific pollutants such as nitrogen and phosphorus compounds, heavy metals and chlorinated compounds. New methods include aerobic, anaerobic and physico-chemical processes in fixed-bed filters and in bioreactors in which the materials and microbes are held in suspension.

The costs of waste water treatment can be reduced by the conversion of wastes into useful products. For example, heavy metals and sulphur compounds can be removed from waste streams of the galvanisation industry by the aid of sulphur metabolising bacteria and reused. Another example is the production of animal feed from the fungal biomass which remains after the production of penicillin. Most anaerobic waste water treatment systems produce useful biogas.

Drinking and Process Water

Abundant supplies of water are vital for modern urban and industrial development. By the turn of this century, it is estimated that two-thirds of the world's nations will be water stressed - using clean water faster than it is replenished in aquifers or rivers. A very important aspect of biotechnology is therefore its potential for the reclamation and purification of waste waters for re-use. Public concern has also increased over the quality of drinking water.

Not only does water need to be recycled in the development of sustainable use of resources, overall quality must also be improved to satisfy consumers. In many agricultural regions of the world, animal wastes and excess fertilisers result in high levels of nitrates in drinking water. Biotechnology has provided successful methods by which these compounds can be removed from processed water before it is delivered to customers.

Air and Waste Gases

Originally, industrial waste gas treatment systems were based on cheap compost-filled filters that removed odours. Such systems still exist. However, slow processing rates and the short life of such filters drove research into better methods such as bioscrubbers, in which the pollutants are washed out using a cell suspension and biotrickling filters, in which the pollutant is degraded by micro-organisms immobilised on an inert matrix and provided with an aqueous nutrient film trickling through the device.

The selection of micro-organisms that are more efficient at metabolising pollutants has also led to better air and gas purifying biofilters. Examples are a bioscrubber based system for the simultaneous removal of nitrogen and sulphur oxides from the flue gas of blastfurnaces which has been developed as an alternative to the classical limestone gypsum process, and the elimination of styrene from the waste gas of polystyrene processing industries by a biofilter containing fungi.

Soil and Land Treatment

Both in situ (in its original place) and ex situ (somewhere else) methods are commercially exploited for the cleanup of soil and the associated groundwater. In situ treatments may include the introduction of micro-organisms (bioaugmentation), ventilation and/or adding nutrient solutions (biostimulation). Ex situ treatment involves removing the soil and groundwater and treating it above ground. The soil may be treated as compost, in soil banks, or in specialised slurry bioreactors. Groundwater is treated in bioreactors and either pumped back into the ground or drained into surface water.

Bioremediation of land (biorestoration) is often cheaper than physical methods and its products are harmless if complete mineralisation takes place. Its action can however, be time-consuming, tying up capital and land.

The in situ bioremediation of the ground under petrol stations has already become common practice but also for chlorinated solvents like tri- and tetrachloroethylene in situ bioremediation is possible. The applicability of in situ bioremediation is and probably will remain dependent on the physical parameters of the soil, mainly its transport properties. Bioremediation using plants is called phytoremediation.

This technique is presently already used to remove metals from contaminated soils and groundwater and is being further explored for the remediation of other pollutants. The combined use of plants and bacteria may also be possible. Certain bacteria live closely associated with the roots of plants and depend on substances excreted by the roots. Such rhizobacteria, whose numbers are much higher than those of other soil bacteria, may be genetically modified to break down pollutants. Research is being conducted to test this hypothesis.

Solid Waste

Domestic solid wastes are a major problem in our consumption society. Their elimination is both costly and warrants constant surveillance in terms of groundwater and air pollution.

Yet, for a major part they are composed of readily biodegradable organics. In this respect, source separated bio-wastes can be converted to a valuable resource by composting or anaerobic digestion. In recent years, both processes have seen remarkable developments in terms of process design and control.

Particularly, anaerobic digestion of solid wastes in high-rate anaerobic digesters has gained increasing public acceptance because it permits the recovery of substantial amounts of high-value biogas together with a high quality stable organic residue and this without giving rise to environmental nuisance. Moreover, anaerobic digestion of mixed solid wastes is under intensive development because in the near future it may be an important step in recycling of solid wastes and constitute an alternative to incineration.

Prevention

Progressively more industrial companies are developing processes with reduced environmental impact responding to the international call for the development of a sustainable society. There is a pervading trend towards less harmful products and processes; away from "end-of-pipe" treatment of waste streams. Biotechnology is pre-eminently suitable to contribute to this trend and it has already done so in many cases, both by the improvement of existing processes and by the development of new ones.

Process Improvement

Many industrial processes have been made more environmentally friendly by the use of enzymes. Enzymes are biological catalysts that are highly efficient and have numerous advantages over non-biological catalysts. They are non-toxic and biodegradable, work best at moderate temperatures and in mild conditions, and have fewer side reactions than traditional methods because they are highly specific. Production methods that employ enzymes are generally not only cleaner and safer compared with other methods, but mostly also more economic in energy and resource consumption.

Their specificity does however mean that it is not always easy to find the appropriate enzyme for a given application. Enzymes are already widely employed in industry and have been for many years. New techniques and approaches to protein design and molecular modelling are enabling researchers to develop novel enzymes active at high temperatures, in non-aqueous solvents and as solids.

Product Innovation

Biotechnology also can help to produce new products which have less impact on the environment than their predecessors. The production of new biomaterials like bioplastics avoids the use of non-renewable resources like fossil fuels. Potatoes normally contain 80% amylopectin but also 20% amylose which is unwanted in many applications.

For the isolation of pure amylopectin large amounts of water and energy are consumed. A Dutch company has developed a genetically modified potato variety which no longer contains amylose and hence can be processed with less impact on the environment.

The use of genetically modified plant varieties that are resistant against insects and/or diseases may considerably diminish the use of pesticides which not only prevents the use of the -mostly non-renewable- raw materials, energy and labour necessary for their production but also will reduce the negative impacts of their residues. Many more of these biotechnological solutions for pollution have been developed.

Future developments may involve things which at this moment seem science-fiction to most people such as the replacement of chemically produced super-fibres by microbially produced spider web silk. One thing that should not be forgotten however is that the increased use of biological systems in industry should be accompanied by adequate training and protection of workers handling these systems, just like in other sections of industry.

DETECTION AND MONITORING

Detection and Monitoring of Pollutants

A wide range of biological methods are already in use to detect pollution incidents and for the continuous monitoring of pollutants. Long established measures include: counting the number of plant, animal and microbial species, counting the numbers of individuals in those species or analysing the levels of oxygen, methane or other compounds in water. More recently, biological detection methods using biosensors and immunoassays have been developed and are now being commercialised. Most biosensors are a combination of biological and electronic devices - often built onto a microchip.

The biological component might be simply an enzyme or antibody, or even a colony of bacteria, a membrane, neural receptor, or an entire organism. Immobilised on a substrate, their properties change in response to some environmental effect in a way that is electronically or optically detectable.

It is then possible to make quantitative measurements of pollutants with extreme precision or to very high sensitivities. The sensors can be designed to be very selective, or sensitive to a broad range of compounds. For example, a wide range of herbicides can be detected in river water using algal-based biosensors; the stresses inflicted on the organisms being measured as changes in the optical properties of the plant's chlorophyll. Microbial biosensors are micro-organisms which produce a reaction upon contact with the substance to be sensed. Usually they produce light but cease to do so upon contact with substances which are toxic to them. Both naturally occurring light emitting microorganisms as well as specially developed ones are used.

Positively acting bacterial biosensors have been constructed which start emitting light upon contact (and subsequent reaction) with a specific pollutant. In the USA such a light emitting bacterium has been approved for the detection of polyhalogenated aromatic hydrocarbons in field tests.

Immunoassays use labelled antibodies (complex proteins produced in biological response to specific agents) and enzymes to measure pollutant levels.

If a pollutant is present, the antibody attaches itself to it; the label making it detectable either through colour change, fluorescence or radioactivity.

Immunoassays of various types have been developed for the continuous, automated and inexpensive monitoring of pesticides such as dieldrin and parathion.

The nature of these techniques, the results of which can be as simple as a colour change, make them particularly suitable for highly sensitive field testing where the time and large equipment needed for more traditional testing is impractical. Their use is however limited to pollutants which can trigger biological antibodies. If the pollutants are too reactive, they will either destroy the antibody or suppress its activity and so also the effectiveness of the test.

Detection and Monitoring of Microorganisms used for Bioremediation

When laboratory grown micro-organisms are inoculated into a bioremediation site (bioaugmentation) it often becomes necessary to monitor their presence and/or multiplication to check the progress of the process. This is especially true and even required when genetically modified micro-organisms are involved. The traditional technique to detect thc presence of micro-organisms in soil is direct plating on selective media.

This is greatly facilitated if the organism contains a marker which can be selected for. Newer techniques include the above mentioned immunological and light-based bioreporter techniques. The spatial distribution of specific microorganisms in a sample can be determined microscopically and non-invasively by using fluorescent in situ hybridisation (FISH) of micro-organisms. The most sensitive and specific technique is the direct isolation and amplification of DNA from soil, which is increasingly being used.

Detection and Monitoring of Ecological Effects

Bioremediation is aimed at improving the quality of the environment by removing pollutants. However, the disappearance of the original pollutant is not the only criterion by which the success of a bioremediation operation is determined. (Even more) toxic metabolites may be produced from the pollutant or the biodegrading bacterium may cause diseases or produce substances that are harmful to useful micro-organisms, plants, animals or humans.

All these negative effects, are of course, excluded as much as possible in advance by getting as familiar as possible with the organism through extensive literature searches and microcosm studies in which the bioremediation process is simulated in the laboratory. To avoid unexpected effects, especially after the release of new member of the eco-system like a genetically modified organism, the monitoring of the ecological effects of a bioremediation operation may be required.

The problem with monitoring ccological effects is what to monitor. Numerous ecological effects are possible but not all of them may be relevant or permanent or even the result of the bioremediation operation. The parameters to be monitored are usually determined case-by-case. Monitoring techniques may include all of those mentioned in the two previous subsections on detection and monitoring.

GENETIC ENGINEERING

Recombinant DNA technology has had amazing repercussions in the last few years. Molecular biologists have mapped entire genomes, many new medicines have been developed and introduced and agriculturists are producing plants with novel types of disease resistance that could not be achieved through conventional breeding.

Several of the previously mentioned examples like the amylose-free potato and the indigo-producing bacterium also involve the use of organisms genetically modified by recombinant DNA technology.

Many enzymes are routinely produced by genetically modified organisms too. Given the overwhelming diversity of species, biomolecules and metabolic pathways on this planet, genetic engineering can in principle be a very powerful tool in creating environmentally friendlier alternatives for products and processes that presently pollute the environment or exhaust its non-renewable resources. Politics, economics and society will ultimately determine which scientific possibilities will become reality.

Nowadays organisms can also be supplemented with additional genetic properties for the biodegradation of specific pollutants if naturally occurring organisms are not able to do that job properly or not quickly enough. By combining different metabolic abilities in the same micro-organism bottlenecks in environmental cleanup may be circumvented. Until now this has not been done on any significant scale. The main reason being the fact that in most cases naturally occurring organisms can be found or selected for, which are able to clean up a polluted site.

Examples have been found where soil bacteria have developed new properties in response to the introduction of xenobiotics (that is, manmade chemicals that are normally not found in nature). In some cases they even appear to have acquired properties from other species. In the USA some genetically modified bacteria have been approved for bioremediation purposes but large scale applications have not yet been reported.

In Europe only controlled field tests have been authorized. Because new organisms can be created by genetic engineering that may never be produced by spontaneous or selection driven evolution, concerns exist about the unpredictability of their possible interactions with the eco-system.

Genetically modified organisms which are properly kept within the confines of their approved production facilities are much less a concern than genetically modified organisms which are meant to be released into the environment like disease resistant plants or soil bacteria for bioremediation.

The possible ecological effects of the latter are even more difficult to evaluate due to the fact that it is well known that soil bacteria frequently exchange genetic material (also between species).

This together with the fact that we know little about the great majority of soil inhabiting bacterial species, makes it almost impossible to predict the fate of every DNA copy of a newly introduced genetic property in a soil bacterium.

If the extra DNA is derived from another soil bacterium, it may on the other hand be reasonable to argue that the genetically modified bacterium might also have evolved spontaneously some day due to the frequent exchange of genetic material in the soil.

LEGISLATION

Regulation to ensure safe application of novel or modified organisms in the environment is important, not least to maintain public confidence. The European Union has two Directives on the contained use of genetically modified micro-organisms, and on the deliberate release of genetically modified organisms into the environment. These have been implemented in the national legislation of most EU Member States. They require that a detailed experimental protocol, including assessment of potential risks, is approved by competent authorities before a genetically modified organism is released into the environment.

The nature and sometimes even the site of the release has to be published in the local press in some countries. After several years' experience using the legislation, the procedures involved are now being revised. Ammendments to clarify and revise Directive 90/219/EEC were published in December 1998. The aim of the European Commission is to maintain the EU's competitiveness globally - both in research and commercial applications- without compromising safety.

PUBLIC OPINION, DIALOGUE AND DEBATE

In spite of the fact that traditional biotechnology already is of great value to bioremediation and modern biotechnology may enhance this even further, there are no recent data on what Europeans specifically think about environmental biotechnology. Generally speaking, Europeans tend to take an "optimistic" view of the developments they expect from modern biotechnology, according to the most recent European Commission public opinion survey which was published in 1997.

Unfortunately this survey did not investigate the attitude of the public towards environmental biotechnology. The only environmentally related question in this survey was whether people believed that modern biotechnology would substantially reduce environmental pollution, which 17 % did. Whether or not this is only wishful thinking remains to be determined. Ultimately, hard proof expessed in the form of improved environmental parameters will be needed for full acceptance of environmental biotechnology.

Conferences, public debates, seminars and round table meetings have been held to bring people from the public, government, environmental organisations, science and industry together to discuss critical issues. These lively debates do not always lead to consensus, but they can provide a fuller appreciation of all the aspects in a particular issue, facilitating a better understanding of the problems involved.

A recent example is the workshop 'How can biotechnology benefit the environment'. Public information aimed at advancing dialogue and debate is provided by many organisations.

A compilation of these can be found in the handbook of information sources which has been published by and can be ordered from the EFB Taskgroup on Public perceptions of Biotechnology.

Environmental biotechnology has a career extending back into the last century. As the need is better appreciated to move towards less destructive patterns of economic activity, while maintaining improvement of social conditions in spite of increasing population, the role of biotechnology grows as a tool for remediation and environmentally sensitive industry. Already, the technology has been proven in a number of areas and future developments promise to widen its scope. Some of the new techniques now under consideration make use of genetically modified organisms designed to deal efficiently with specific tasks.

As with all situations where there is to be a release of new technology into the environment, concerns exist. There is a potential for biotechnology to make a further major contribution to protection and remediation of the environment. Hence biotechnology is well positioned to contribute to the development of a more sustainable society. As we move into the next millennium this will become even more vitally important as populations, urbanisation and industrialisation will continue to climb.

More Sustainable Industrial Processes Through the use of Enzymes

The leather processing industry has introduced enzymes to replace harsh chemicals traditionally used for cleaning the hide. In textile production, enzymes have superseded chemicals for bleaching, including the "stone washing" of jeans. Chlorine consumption by the pulp and paper industry may soon also be reduced considerably by the use of enzymes. The grease and protein digesting enzymes in washing powders significantly reduce the quantity of detergents needed for a given washing effect.

They also mean that the washing temperature can be reduced. Lowering the temperature 20°C saves more than a third of the energy used by the machine. Since in many Western European countries up to 5% of household energy consumption was used for washing, these molecules have made a significant contribution to energy conservation.

Biotechnological Solutions for Pollution

Pigs and chickens cannot utilize phosphate from phytate in their feed, which therefore ends up in their manure. By adding the enzyme phytase to their feed the amount of phosphate which is excreted by these animals can be reduced by more than 30 %. In South Africa bacteria are used for the isolation of gold from gold-ore. This so-called biomining saves an enormous amount of smelting energy and generates much less waste.

The chemical production of indigo, the dye which is used for blue jeans, takes eight steps, the use of very toxic chemicals and special protection measures for the process operators and the environment. The biotechnological production of indigo, which uses a genetically modified bacterium containing the right enzymes, takes only three steps, proceeds in water, uses simple raw materials like sugar and salts and generates only indigo, carbon dioxide and biomass which is biodegradable.

2

Fundamental of Microbiology

Microbiology is the study of microorganisms, which are microscopic, unicellular, and cell-cluster organisms. This includes eukaryotes such as fungi and protists, and prokaryotes. Viruses and prions, though not strictly classed as living organisms, are also studied. Microbiology typically includes the study of the immune system, or Immunology. Generally, immune systems interact with pathogenic microbes; these two disciplines often intersect which is why many colleges offer a paired degree such as "Microbiology and Immunology".

Microbiology is a broad term which includes virology, mycology, parasitology, bacteriology and other branches. A microbiologist is a specialist in microbiology and these other topics. Microbiology is researched actively, and the field is advancing continually. It is estimated only about one per cent of all of the microbe species on Earth have been studied. Although microbes were directly observed over three hundred years ago, the field of microbiology can be said to be in its infancy relative to older biological disciplines such as zoology and botany.

HISTORY

ANCIENT

The existence of microorganisms was hypothesized for many centuries before their actual discovery. The existence of unseen microbiological life was postulated by Jainism which is based on Mahavira's teachings as early as 6th century BCE.. Paul Dundas notes that Mahavira asserted existence of unseen microbiological creatures living in earth, water, air and fire. Jain scriptures also describe nigodas which are sub-microscopic creatures living in large clusters and having a very short life and are said to pervade each and every part of the universe, even in tissues of plants and flesh of animals.

The Roman Marcus Terentius Varro made references to microbes when he warned against locating a homestead in the vicinity of swamps "because there are bred certain minute creatures which cannot be seen by the eyes, which float in the air and enter the body through the mouth and nose and there cause serious diseases."

In 1546 Girolamo Fracastoro proposed that epidemic diseases were caused by transferable seedlike entities that could transmit infection by direct or indirect contact, or even without contact over long distances. However, early claims about the existence of microorganisms were speculative, and not based on any data or observation. Actual observation and discovery of microbes had to await the invention of the microscope in the 17th century.

MODERN

in 1676, Antonie van Leeuwenhoek observed bacteria and other microorganisms, using a single-lens microscope of his own design. While Van Leeuwenhoek is often cited as the first to observe microbes, Robert Hooke made the first recorded microscopic observation, of the fruiting bodies of molds, in 1665.. The first observation of microbes using a microscope is generally credited to the Dutch draper and haberdasher, Antonie van Leeuwenhoek, who lived for most of his life in Delft, Holland.

It has, however, been suggested that a Jesuit priest called Athanasius Kircher was the first to observe microorganisms. He was among the first to design magic lanterns for projection purposes, so he must have been well acquainted with the properties of lenses. One of his books, which reads in translation–'Concerning the wonderful structure of things in nature, investigated by Microscope.

Here, he wrote 'who would believe that vinegar and milk abound with an innumerable multitude of worms.' He also noted that putrid material is full of innumerable creeping animalculae. These observations antedate Robert Hooke's Micrographia by nearly 20 years and were published some 29 years before van Leeuwenhoek saw protozoa and 37 years before he described having seen bacteria. The field of bacteriology was founded in the 19th century by Ferdinand Cohn, a botanist whose studies on algae and photosynthetic bacteria led him to describe several bacteria including Bacillus and Beggiatoa. Cohn was also the first to formulate a scheme for the taxonomic classification of bacteria and discover spores.

Louis Pasteur and Robert Koch were contemporaries of Cohn's and are often considered to be the father of Microbiology and medical microbiology, respectively. Pasteur is most famous for his series of experiments designed to disprove the then widely held theory of spontaneous generation, thereby solidifying microbiology's identity as a biological science. Pasteur also designed methods for food preservation and vaccines against several diseases such as anthrax, fowl cholera and rabies. Koch is best known for his contributions to the germ theory of disease, proving that specific diseases were caused by specific pathogenic microorganisms. He developed a series of criteria that have become known as the Koch's postulates. Koch was one of the first scientists to focus on the isolation of bacteria in pure culture resulting in his description of several novel bacteria including Mycobacterium tuberculosis, the causative agent of tuberculosis.

While Pasteur and Koch are often considered the founders of microbiology, their work did not accurately reflect the true diversity of the microbial world because of their exclusive focus on microorganisms having direct medical relevance. It was not until the late 19th century and the work of Martinus Beijerinck and Sergei Winogradsky, the founders of general microbiology, that the true breadth of microbiology was revealed. Beijerinck made two major contributions to microbiology: the discovery of viruses and the development of enrichment culture techniques. While his work on the Tobacco Mosaic Virus established the basic principles of virology, it was his development of enrichment culturing that had the most immediate impact on microbiology by allowing for the cultivation of a wide range of microbes with wildly different physiologies. Winogradsky was the first to develop the concept of chemolithotrophy and to thereby reveal the essential role played by microorganisms in geochemical processes. He was responsible for the first isolation and description of both nitrifying and nitrogen-fixing bacteria.

FIELDS

The field of microbiology can be generally divided into several subdisciplines:

- *Aeromicrobiology*: The study of airborne microorganisms.
- *Agricultural microbiology*: The study of agriculturaly important microorganisms.
- *Cellular microbiology*: A discipline bridging microbiology and cell biology.
- *Environmental microbiology*: The study of the function and diversity of microbes in their natural environments. Includes the study of microbial ecology, microbially-mediated nutrient cycling, geomicrobiology, microbial diversity and bioremediation. Characterization of key bacterial habitats such as the rhizosphere and phyllosphere, soil and groundwater ecosystems, open oceans or extreme environments.
- *Evolutionary microbiology*: The study of the evolution of microbes. Includes the study of bacterial systematics and taxonomy.
- *Food microbiology*: The study of microorganisms causing food spoilage and foodborne illness. Using microorganisms to produce foods, for example by fermentation.
- *Generation microbiology*: The study of those microorganisms that have the same characters as their parents.

- *Industrial microbiology*: The exploitation of microbes for use in industrial processes. Examples include industrial fermentation and wastewater treatment. Closely linked to the biotechnology industry. This field also includes brewing, an important application of microbiology.
- *Medical microbiology*: The study of the pathogenic microbes and the role of microbes in human illness. Includes the study of microbial pathogenesis and epidemiology and is related to the study of disease pathology and immunology.
- *Microbial genetics*: The study of how genes are organized and regulated in microbes in relation to their cellular functions. Closely related to the field of molecular biology.
- *Microbial physiology*: The study of how the microbial cell functions biochemically. Includes the study of microbial growth, microbial metabolism and microbial cell structure.
- *Nano microbiology*: The study of those microorganisms at nano level.
- *Pharmaceutical microbiology*: the study of microorganisms causing pharmaceutical contamination and spoil
- *Soil Microbiology*: The study of those microorganisms that are found in soil.
- *Veterinary microbiology*: The study of the role in microbes in veterinary medicine or animal taxonomy.
- *Water microbiology*: The study of those microorganisms that are found in water.

BENEFITS

Whilst there are undoubtedly some who fear all microbes due to the association of some microbes with various human illnesses, many microbes are also responsible for numerous beneficial processes such as industrial fermentation, antibiotic production and as vehicles for cloning in more complex organisms such as plants. Scientists have also exploited their knowledge of microbes to produce biotechnologically important enzymes such as Taq polymerase, reporter genes for use in other genetic systems and novel molecular biology techniques such as the yeast two-hybrid system.

Bacteria can be used for the industrial production of amino acids. Corynebacterium glutamicum is one of the most important bacterial species with an annual production of more than two million tons of amino acids, mainly L-glutamate and L-lysine. A variety of biopolymers, such as polysaccharides, polyesters, and polyamides, are produced by microorganisms.

Microorganisms are used for the biotechnological production of biopolymers with tailored properties suitable for high-value medical application such as tissue engineering and drug delivery. Microorganisms are used for the biosynthesis of xanthan, alginate, cellulose, cyanophycin, poly, levan, hyaluronic acid, organic acids, oligosaccharides and polysaccharide, and polyhydroxyalkanoates. Microorganisms are beneficial for microbial biodegradation or bioremediation of domestic, agricultural and industrial wastes and subsurface pollution in soils, sediments and marine environments.

The ability of each microorganism to degrade toxic waste depends on the nature of each contaminant. Since sites typically have multiple pollutant types, the most effective approach to microbial biodegradation is to use a mixture of bacterial species and strains, each specific to the biodegradation of one or more types of contaminants. There are also various claims concerning the contributions to human and animal health by consuming probiotics and/or prebiotics.

Recent research has suggested that microorganisms could be useful in the treatment of cancer. Various strains of non-pathogenic clostridia can infiltrate and replicate within solid tumors. Clostridial vectors can be safely administered and their potential to deliver therapeutic proteins has been demonstrated in a variety of preclinical models.

MICROORGANISM

A microorganism or microbe is an organism that is unicellular or lives in a colony of cellular organisms. The study of microorganisms is called microbiology, a subject that began with Anton van Leeuwenhoek's discovery of microorganisms in 1675, using a microscope of his own design. Microorganisms are very diverse; they include bacteria, fungi, archaea, and protists; microscopic plants; and animals such as plankton and the planarian. Some microbiologists also include viruses, but others consider these as non-living.

Most microorganisms are unicellular but this is not universal, since some multicellular organisms are microscopic, while some unicellular protists and bacteria, like Thiomargarita namibiensis, are macroscopic and visible to the naked eye. Microorganisms live in all parts of the biosphere where there is liquid water, including soil, hot springs, on the ocean floor, high in the atmosphere and deep inside rocks within the Earth's crust.

Microorganisms are critical to nutrient recycling in ecosystems as they act as decomposers. As some microorganisms can fix nitrogen, they are a vital part of the nitrogen cycle, and recent studies indicate that airborne microbes may play a role in precipitation and weather. Microbes are also exploited by people in biotechnology, both in traditional food and beverage preparation, and in modern technologies based on genetic engineering. However, pathogenic microbes are harmful, since they invade and grow within other organisms, causing diseases that kill people, other animals and plants.

HISTORY

Evolution

Single-celled microorganisms were the first forms of life to develop on Earth, approximately 3–4 billion years ago. Further evolution was slow, and for about 3 billion years in the Precambrian eon, all organisms were microscopic. So, for most of the history of life on Earth the only forms of life were microorganisms. Bacteria, algae and fungi have been identified in amber that is 220 million years old, which shows that the morphology of microorganisms has changed little since the Triassic period.

Most microorganisms can reproduce rapidly and microbes such as bacteria can also freely exchange genes by conjugation, transformation and transduction between widely-divergent species. This horizontal gene transfer, coupled with a high mutation rate and many other means of genetic variation, allows microorganisms to swiftly evolve to survive in new environments and respond to environmental stresses. This rapid evolution is important in medicine, as it has led to the recent development of 'super-bugs'—pathogenic bacteria that are resistant to modern antibiotics.

Pre-microbiology

The possibility that microorganisms exist was discussed for many centuries before their actual discovery in the 17th century. The existence of unseen microbiological life was postulated by Jainism which is based on Mahavira's teachings as early as 6th century BCE. Paul Dundas notes that Mahavira asserted existence of unseen microbiological creatures living in earth, water, air and fire.

Jain scriptures also describe nigodas which are sub-microscopic creatures living in large clusters and having a very short life and are said to pervade each and every part of universe, even in tissues of plants and flesh of animals. However, the earliest known idea to indicate the possibility of diseases spreading by yet unseen organisms was that of the Roman scholar Marcus Terentius Varro in a 1st century BC book titled On Agriculture in which he warns against locating a homestead near swamps:

- ...and because there are bred certain minute creatures which cannot be seen by the eyes, which float in the air and enter the body through the mouth and nose and there cause serious diseases.

In The Canon of Medicine, Avicenna stated that bodily secretion is contaminated by foul foreign earthly bodies before being infected. He also hypothesized that tuberculosis and other diseases might be contagious, *i.e.* that they were infectious diseases, and used quarantine to limit their spread. When the Black Death bubonic plague reached Andalusia in Spain, in the 14th century, Ibn Khatima wrote that infectious diseases were caused by contagious "minute bodies" that enter the human body. Later, in 1546, Girolamo Fracastoro proposed that epidemic diseases were caused by transferable seedlike entities that could transmit infection by direct or indirect contact, or even without contact over long distances. All these early claims about the existence of microorganisms were speculative and were not based on any data or science. Microorganisms were neither proven, observed, nor correctly and accurately described until the 17th century. The reason for this was that all these early studies lacked the microscope.

History of Microorganisms' Discovery

Anton van Leeuwenhoek was one of the first people to observe microorganisms, using a microscope of his own design, and made one of the most important contributions to biology. Robert Hooke was the first to use a microscope to observe living things; his 1665 book Micrographia contained descriptions of plant cells. Before Leeuwenhoek's discovery of microorganisms in 1675, it had been a mystery why grapes could be turned into wine, milk into cheese, or why food would spoil. Leeuwenhoek did not make the connection between these processes and microorganisms, but using a microscope, he did establish that there were forms of life that were not visible to the naked eye. Leeuwenhoek's discovery, along with subsequent observations by Lazzaro Spallanzani and Louis Pasteur, ended the long-held belief that life spontaneously appeared from non-living substances during the process of spoilage.

Lazzaro Spallanzani found that boiling broth would sterilise it and kill any microorganisms in it. He also found that new microorganisms could only settle in a broth if the broth was exposed to the air. Louis Pasteur expanded upon Spallanzani's findings by exposing boiled broths to the air, in vessels that contained a filter to prevent all particles from passing through to the growth medium, and also in vessels with no filter at all, with air being admitted via a curved tube that would not allow dust particles to come in contact with the broth. By boiling the broth beforehand, Pasteur ensured that no microorganisms survived within the broths at the beginning of his experiment. Nothing grew in the broths in the course of Pasteur's experiment. This meant that the living organisms that grew in such broths came from outside, as spores on dust, rather than spontaneously generated within the broth. Thus, Pasteur dealt the death blow to the theory of spontaneous generation and supported germ theory.

In 1876, Robert Koch established that microbes can cause disease. He found that the blood of cattle who were infected with anthrax always had large numbers of Bacillus anthracis. Koch found that he could transmit anthrax from one animal to another by taking a small sample of blood from the infected animal and injecting it into a healthy one, and this caused the healthy animal to become sick. He also found that he could grow the bacteria in a nutrient broth, then inject it into a healthy animal, and cause illness.

Based on these experiments, he devised criteria for establishing a causal link between a microbe and a disease and these are now known as Koch's postulates. Although these postulates cannot be applied in all cases, they do retain historical importance to the development of scientific thought and are still being used today.

CLASSIFICATION AND STRUCTURE

Microorganisms can be found almost anywhere in the taxonomic organization of life on the planet. Bacteria and archaea are almost always microscopic, while a number of eukaryotes are also microscopic, including most protists, some fungi, as well as some animals and plants. Viruses are generally regarded as not living and therefore are not microbes, although the field of microbiology also encompasses the study of viruses.

Prokaryotes

Prokaryotes are organisms that lack a cell nucleus and the other membrane bound organelles. They are almost always unicellular, although some species such as myxobacteria can aggregate into complex structures as part of their life cycle.Consisting of two domains, bacteria and archaea, the prokaryotes are the most diverse and abundant group of organisms on Earth and inhabit practically all environments where some liquid water is available and the temperature is below +140 °C. They are found in sea water, soil, air, animals' gastrointestinal tracts, hot springs and even deep beneath the Earth's crust in rocks. Practically all surfaces which have not been specially sterilized are covered by prokaryotes. The number of prokaryotes on Earth is estimated to be around five million trillion trillion, or 5 × 10, accounting for at least half the biomass on Earth.

Bacteria

Bacteria are practically all invisible to the naked eye, with a few extremely rare exceptions, such as Thiomargarita namibiensis. They lack membrane-bound organelles, and can function and reproduce as individual cells, but often aggregate in multicellular colonies. Their genome is usually a single loop of DNA, although they can also harbor small pieces of DNA called plasmids. These plasmids can be transferred between cells through bacterial conjugation. Bacteria are surrounded by a cell wall, which provides strength and rigidity to their cells. They reproduce by binary fission or sometimes by budding, but do not undergo sexual reproduction. Some species form extraordinarily resilient spores, but for bacteria this is a mechanism for survival, not reproduction. Under optimal conditions bacteria can grow extremely rapidly and can double as quickly as every 10 minutes.

Archaea

Archaea are also single-celled organisms that lack nuclei. In the past, the differences between bacteria and archaea were not recognised and archaea were classified with bacteria as part of the kingdom Monera. However, in 1990 the microbiologist Carl Woese proposed the three-domain system that divided living things into bacteria, archaea and eukaryotes. Archaea differ from bacteria in both their genetics and biochemistry.

For example, while bacterial cell membranes are made from phosphoglycerides with ester bonds, archaean membranes are made of ether lipids. Archaea were originally described in extreme environments, such as hot springs, but have since been found in all types of habitats. Only now are scientists beginning to realise how common archaea are in the environment, with crenarchaeota being the most common form of life in the ocean, dominating ecosystems below 150 m in depth. These organisms are also common in soil and play a vital role in ammonia oxidation.

Eukaryotes

Most living things which are visible to the naked eye in their adult form are eukaryotes, including humans. However, a large number of eukaryotes are also microorganisms. Unlike bacteria and archaea, eukaryotes contain organelles such as the cell nucleus, the Golgi apparatus and mitochondria in their cells. The nucleus is an organelle which houses the DNA that makes up a cell's genome. DNA itself is arranged in complex chromosomes. Mitochondria are organelles vital in metabolism as they are the site of the citric acid cycle and oxidative phosphorylation. They evolved from symbiotic bacteria and retain a remnant genome. Like bacteria, plant cells have cell walls, and contain organelles such as chloroplasts in addition to the organelles in other eukaryotes. Chloroplasts produce energy from light by photosynthesis, and were also originally symbiotic bacteria.

Unicellular eukaryotes are those eukaryotic organisms that consist of a single cell throughout their life cycle. This qualification is significant since most multicellular eukaryotes consist of a single cell called a zygote at the beginning of their life cycles. Microbial eukaryotes can be either haploid or diploid, and some organisms have multiple cell nuclei. However, not all microorganisms are unicellular as some microscopic eukaryotes are made from multiple cells. Of eukaryotic groups, the protists are most commonly unicellular and microscopic. This is a highly diverse group of organisms that are not easy to classify. Several algae species are multicellular protists, and slime molds have unique life cycles that involve switching between unicellular, colonial, and multicellular forms. The number of species of protozoa is uncertain, since we may have identified only a small proportion of the diversity in this group of organisms.

Animals

Mostly animals are multicellular, but some are too small to be seen by the naked eye. Microscopic arthropods include dust mites and spider mites. Microscopic crustaceans include copepods and the cladocera, while many nematodes are too small to be seen with the naked eye. Another particularly common group of microscopic animals are the rotifers, which are filter feeders that are usually found in fresh water. Micro-animals reproduce both sexually and asexually and may reach new habitats as eggs that survive harsh environments that would kill the adult animal. However, some simple animals, such as rotifers and nematodes, can dry out completely and remain dormant for long periods of time.

Fungi

The fungi have several unicellular species, such as baker's yeast and fission yeast. Some fungi, such as the pathogenic yeast Candida albicans, can undergo phenotypic switching and grow as single cells in some environments, and filamentous hyphae in others. Fungi reproduce both asexually, by budding or binary fission, as well by producing spores, which are called conidia when produced asexually, or basidiospores when produced sexually.

Plants

The green algae are a large group of photosynthetic eukaryotes that include many microscopic organisms. Although some green algae are classified as protists, others such as charophyta are classified with embryophyte plants, which are the most familiar group of land plants. Algae can grow as single cells, or in long chains of cells. The green algae include unicellular and colonial flagellates, usually but not always with two flagella per cell, as well as various colonial, coccoid, and filamentous forms. In the Charales, which are the algae most closely related to higher plants, cells differentiate into several distinct tissues within the organism. There are about 6000 species of green algae.

HABITATS AND ECOLOGY

Microorganisms are found in almost every habitat present in nature. Even in hostile environments such as the poles, deserts, geysers, rocks, and the deep sea. Some types of microorganisms have adapted to the extreme conditions and sustained colonies; these organisms are known as extremophiles. Extremophiles have been isolated from rocks as much as 7 kilometres below the Earth's surface, and it has been suggested that the amount of living organisms below the Earth's surface may be comparable with the amount of life on or above the surface. Extremophiles have been known to survive for a prolonged time in a vacuum, and can be highly resistant to radiation, which may even allow them to survive in space.

Many types of microorganisms have intimate symbiotic relationships with other larger organisms; some of which are mutually beneficial while others can be damaging to the host organism. If microorganisms can cause disease in a host they are known as pathogens.

Extremophiles

Extremophiles are microorganisms which have adapted so that they can survive and even thrive in conditions that are normally fatal to most life-forms.

For example, some species have been found in the following extreme environments:

- *Temperature*: As high as 130 °C, as low as –17 °C
- *Acidity/alkalinity*: Less than pH 0, up to pH 11.5
- *Salinity*: Up to saturation
- *Pressure*: Up to 1,000-2,000 atm, down to 0 atm
- *Radiation*: Up to 5kGy

Extremophiles are significant in different ways. They extend terrestrial life into much of the Earth's hydrosphere, crust and atmosphere, their specific evolutionary adaptation mechanisms to their extreme environment can be exploited in bio-technology, and their very existence under such extreme conditions increases the potential for extraterrestrial life.

Soil Microbes

The nitrogen cycle in soils depends on the fixation of atmospheric nitrogen. One way this can occur is in the nodules in the roots of legumes that contain symbiotic bacteria of the genera Rhizobium, Mesorhizobium, Sinorhizobium, Bradyrhizobium, and Azorhizobium.

Symbiotic Microbes

Symbiotic microbes such as fungi and algae form an association in lichen. Certain fungi form mycorrhizal symbioses with trees that increase the supply of nutrients to the tree.

IMPORTANCE

Microorganisms are vital to humans and the environment, as they participate in the Earth's element cycles such as the carbon cycle and nitrogen cycle, as well as fulfilling other vital roles in virtually all ecosystems, such as recycling other organisms' dead remains and waste products through decomposition. Microbes also have an important place in most higher-order multicellular organisms as symbionts. Many blame the failure of Biosphere 2 on an improper balance of microbes.

- *Use in food*: Microorganisms are used in brewing, winemaking, baking, pickling and other food-making processes. They are also used to control the fermentation process in the production of cultured dairy products such as yogurt and cheese. The cultures also provide flavour and aroma, and inhibit undesirable organisms.
- *Use in water treatment*: Specially-cultured microbes are used in the biological treatment of sewage and industrial waste effluent, a process known as bioaugmentation.
- *Use in energy*: Microbes are used in fermentation to produce ethanol, and in biogas reactors to produce methane. Scientists are researching the use of algae to produce liquid fuels, and bacteria to convert various forms of agricultural and urban waste into usable fuels.
- *Use in science*: Microbes are also essential tools in biotechnology, biochemistry, genetics, and molecular biology. The yeasts and fission yeast are important model organisms in science, since they are simple eukaryotes that can be grown rapidly in large numbers and are easily manipulated. They are particularly valuable in genetics, genomics and proteomics. Microbes can be harnessed for uses such as creating steroids and treating skin diseases. Scientists are also considering using microbes for living fuel cells, and as a solution for pollution.
- *Use in warfare*: In the Middle Ages, diseased corpses were thrown into castles during sieges using catapults or other siege engines. Individuals near the corpses were exposed to the deadly pathogen and were likely to spread that pathogen to others.

IMPORTANCE IN HUMAN HEALTH

- *Human digestion*: Microorganisms can form an endosymbiotic relationship with other, larger organisms. For example, the bacteria that live within the human digestive system contribute to gut immunity, synthesise vitamins such as folic acid and biotin, and ferment complex indigestible carbohydrates.
- *Diseases and immunology*: Microorganisms are the cause of many infectious diseases. The organisms involved include pathogenic bacteria, causing diseases such as plague, tuberculosis and anthrax; protozoa, causing diseases such as malaria, sleeping sickness and toxoplasmosis; and also fungi causing diseases such as ringworm, candidiasis or histoplasmosis. However, other diseases such as

influenza, yellow fever or AIDS are caused by pathogenic viruses, which are not usually classified as living organisms and are not therefore microorganisms by the strict definition. As of 2007, no clear examples of archaean pathogens are known, although a relationship has been proposed between the presence of some methanogens and human periodontal disease.

HYGIENE

Hygiene is the avoidance of infection or food spoiling by eliminating microorganisms from the surroundings. As microorganisms, particularly bacteria, are found practically everywhere, this means in most cases the reduction of harmful microorganisms to acceptable levels.

However, in some cases it is required that an object or substance be completely sterile, *i.e.* devoid of all living entities and viruses. A good example of this is a hypodermic needle. In food preparation microorganisms are reduced by preservation methods, clean utensils used in preparation, short storage periods or by cool temperatures. If complete sterility is needed, the two most common methods are irradiation and the use of an autoclave, which resembles a pressure cooker.

There are several methods for investigating the level of hygiene in a sample of food, drinking water, equipment etc. Water samples can be filtrated through an extremely fine filter.

This filter is then placed in a nutrient medium. Microorganisms on the filter then grow to form a visible colony. Harmful microorganisms can be detected in food by placing a sample in a nutrient broth designed to enrich the organisms in question. Various methods, such as selective media or PCR, can then be used for detection.

The hygiene of hard surfaces, such as cooking pots, can be tested by touching them with a solid piece of nutrient medium and then allowing the microorganisms to grow on it. There are no conditions where all microorganisms would grow, and therefore often several different methods are needed. For example, a food sample might be analysed on three different nutrient mediums designed to indicate the presence of "total" bacteria, molds and coliform bacteria.

THE SPECTRUM OF MICROBIOLOGY

Like all other living things, microorganisms are placed into a system of classification. Classification highlights characteristics that are common among certain groups while providing order to the variety of living things. The science of classification is known as taxonomy, and taxon is an alternative expression for a classification category. Taxonomy displays the unity and diversity among living things, including microorganisms. Among the first taxonomists was Carolus

Linnaeus. In the 1750s and 1760s, Linnaeus classified all known plants and animals of that period and set down the rules for nomenclature.

CLASSIFICATION SCHEMES

The fundamental rank of the classification as set down by Linnaeus is the species. For organisms such as animals and plants, a species is defined as a population of individuals that breed among themselves. For microorganisms, a species is defined as a group of organisms that are 70 per cent similar from a biochemical standpoint.

In the classification scheme, various species are grouped together to form a genus. Among the bacteria, for example, the species *Shigella boydii* and *Shigella flexneri* are in the genus *Shigella* because the organisms are at least 70 per cent similar. Various genera are then grouped as a family because of similarities, and various families are placed together in an order. Continuing the classification scheme, a number of orders are grouped as a class, and several classes are categorized in a single phylum or division.

The various phyla or divisions are placed in the broadest classification entry, the kingdom. Numerous criteria are used in establishing a species and in placing species together in broader classification categories. Morphology and structure are considered, as well as cellular features, biochemical properties, and genetic characteristics. In addition, the antibodies that an organism elicits in the human body are a defining property. The nutritional format is considered, as are staining characteristics.

PROKARYOTES AND EUKARYOTES

Because of their characteristics, microorganisms join all other living organisms in two major groups of organisms: prokaryotes and eukaryotes. Bacteria are prokaryotes because of their cellular properties, while other microorganisms such as fungi, protozoa, and unicellular algae are eukaryotes. Viruses are neither prokaryotes nor eukaryotes because of their simplicity and unique characteristics.

THE FIVE KINGDOMS

The generally accepted classification of living things was devised by Robert Whittaker of Cornell University in 1969. Whittaker suggested a five-kingdom classification.

The first of the five kingdoms is Monera. Prokaryotes, such as bacteria and cyanobacteria are in this kingdom; the second kingdom, Protista, includes protozoa, unicellular algae, and slime molds, all of which are eukaryotes and single-celled; in the third kingdom, Fungi, are the molds, mushrooms, and yeasts. These organisms are eukaryotes that absorb simple nutrients from the soil. The remaining two kingdoms are Plantae and Animalia.

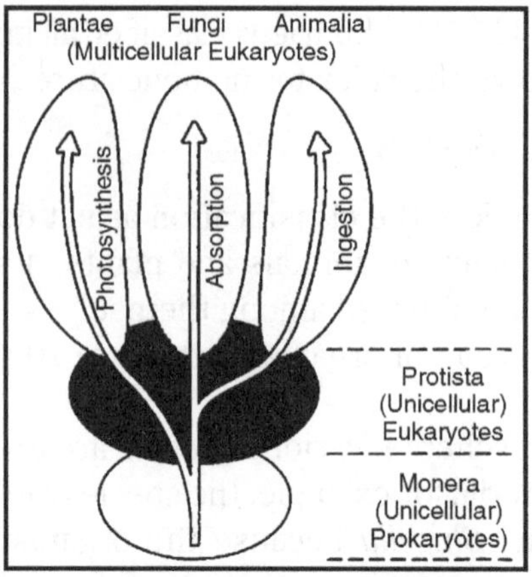

Fig. The Five Kingdoms of Living things Showing the Position of Microorganisms Relative to Other Organisms.

BRIEF DESCRIPTIONS OF MICROORGANISMS

Bacteria are relatively simple, prokaryotic organisms whose cells lack a nucleus or nuclear membrane. The bacteria may appear as rods, spheres, or spirals. Bacteria reproduce by binary fission, have unique constituents in their cell walls, and exist in most environments on earth. For instance, they live at temperatures ranging from 0° to 100°C and in conditions that are oxygen rich or oxygen free. A microscope is necessary to see and study them. Fungi are eukaryotic microorganisms that include multicellular molds and unicellular yeasts.

The yeasts are slightly larger than bacteria and are used in alcoholic fermentations and bread making. Certain yeasts such as *Candida albicans* are pathogenic. Molds are filamentous, branched fungi that use spores for reproduction. The fungi prefer acidic environments, and most live at room temperature under oxygen-rich conditions. The common mushroom is a fungus.

Protozoa are eukaryotic, unicellular organisms. Motion is a characteristic associated with many species, and the protozoa can be classified just as to how they move: Some protozoa use flagella, others use cilia, and others use pseudopodia.

Certain species are nonmotile. Protozoa exist in an infinite variety of shapes because they have no cell walls. Many species cause such human diseases as malaria, sleeping sickness, dysentery, and toxoplasmosis. The term algae implies a variety of plantlike organisms. In microbiology, several types of single-celled algae are important. Examples are the diatoms and dinoflagellates that inhabit the oceans and are found at the bases of marine food chains. Most algae capture sunlight and transform it to the chemical energy of carbohydrates in the process of photosynthesis.

Viruses are ultramicroscopic bits of genetic material enclosed in a protein shell and, sometimes, a membranous envelope. Viruses have no metabolism; therefore, it is difficult to use drugs to interfere with their structures or activities. Viruses multiply in living cells and use the chemical machinery of the cells for their own purpose. Often, they destroy the cells in the process of replicating.

NOMENCLATURE OF MICROORGANISMS

The system for naming all living things, established by Linnaeus, is also applied to microorganisms. In this system, all organisms are placed into a classification system, and each organism is given a binomial name. The binomial name consists of two names. The first name is the genus to which the organism belongs.

The second name is a modifying adjective called the species modifier. In writing the binomial name, the first letter of the genus name is capitalized, and the remainder of the genus name and the complete species modifier are written in lowercase letters. The entire binomial name is either italicized or underlined. It can be abbreviated by using the first letter of the genus name and the full species modifier. An example of a microbial name is *Escherichia coli*, the bacterial rod found in the human intestine. The name is abbreviated *E. coli*.

STEM CELLS

Tissue engineering is an interdisciplinary field that applies the principles and methods of bioengineering, material science, and life sciences toward the assembly of biologic substitutes that will restore, maintain, and improve tissue functions following damage either by disease or traumatic processes. The general principles of tissue engineering involve combining living cells with a natural/synthetic support or scaffold to build a three-dimensional living construct that is functionally, structurally and mechanically equal to or better than the tissue that is to be replaced.

The development of such a construct requires a careful selection of four key materials:

- Scaffold,
- Growth factors,
- Extra cellular matrix,
- Cells.

Much progress has been made in the field of tissue engineering, but further work toward organ and tissue replacement is necessary. The optimal cell source, scaffold design, and in vitro bioreactors, the use and development of micro fabrication technology to create vascularized tissues and organs are still being investigated.

The search for and use of an appropriate multipotent or pluripotent stem cell in tissue engineering is an emerging concept. Certainly, many areas of stem cell research and their potential clinical applications are associated with controversies; therefore, it is important to address the ethical, legal, and social issues early.

This paper will provide an overview of tissue engineering and stem cells, and describe the current progress with stem cell research in tissue engineering, the potential implications on medical treatment and the economic impact with the passage of Proposition.

Artificial transplantation or transplanted organs is a successful therapy for otherwise incurable end-stage diseases or tissue loss. However, such interventions are challenged by organ shortage, the necessity of lifelong immunosuppression and its potential for serious complications.

Scaffold materials are three-dimensional tissue structures that guide the organization, growth and differentiation of cells. Scaffolds must be biocompatible and designed to meet both nutritional and biological needs for the specific cell population.

Growth factors are soluble peptides capable of binding cellular receptors and producing either a permissive or preventive cellular response toward differentiation and/or proliferation of tissue.

ECM must be capable of providing the optimal conditions for cell adhesion, growth, and differentiation within the construct by creating a system capable of controlling environmental factors such as pH, temperature, oxygen tension, and mechanical forces.

These conditions are determined by the particular cell lines and the properties of the scaffold. Finally, the development of a viable construct involves a suitable supply of cells that are ideally nonimmunogenic, highly proliferative, easy to harvest, and have the ability to differentiate into a variety of cell types with specialized functions.

There are two primary methods to harvest cells and culture. In cases where direct harvest is not feasible, as seen in many patients with extensive end-stage organ failure or cells with limited proliferative capacity in culture, stem cells are envisioned as being an alternative source of cells. Despite the growing interest in embryonic stem cell research, progress has been hampered by ethical and legislative debates.

The political, ethical, and religious opposition toward embryonic stem cell research, which primarily uses discarded nontransferred human embryos for their derivation, have biased research toward adult stem cells and severely restricted federal funding in the United States. In the mist of such controversies, California has become the leading state to support stem cell research by passing Proposition or the California Stem Cell Research and Cures Initiative. It promises to provide California a possible solution to bridge critical research funding gaps by committing $3 billion for stem cell research and to potentially accelerate the understanding of their therapeutic potential.

It does not however support human reproductive cloning. The purpose of this paper is to provide an overview of tissue engineering and stem cells, and describe the current progress with stem cell research in tissue engineering and the potential implications on medical.

FUNDAMENTALS OF TISSUE ENGINEERING

In 1933, the concept of tissue engineering was first introduced when mouse tumor cells demonstrated survival when encased in a biocompatible polymer membrane and implanted into the abdominal cavity of chick embryos. Several decades later, Chick et al., demonstrated that pancreatic beta cells from neonatal rats, cultured on synthetic capillaries and perfused with medium, released insulin in response to changes in glucose concentration.

In the early 1980s, Burke et al., successfully created artificial skin with fibroblasts seeded onto collagen scaffolds for the treatment of extensive burn injury. Clinically, this is still being utilized today.

Efforts are now being undertaken for engineering a variety of tissue and organ types with an emphasis on the application of stem cells.

Ultimately, the goal of tissue engineering is to regenerate tissues and restore organ function through cell implantation and matrix incorporation into the patient.

Current approaches to tissue engineering can be stratified into substitutive, histioconductive, and histioinductive. Substitutive approaches (ex vivo) are essentially whole organ replacement, whereas histioconductive approaches (ex vivo) involve the replacement of missing or damaged parts of an organ tissue with ex-vivo constructs.

In contrast, histioinductive approaches facilitate self-repair and may involve gene therapy using DNA delivery via plasmid vectors or growth factors. A number of criteria must be satisfied in order to achieve effective, long-lasting repair of damaged tissues.

- An adequate number of cells must be produced to fill the defect.
- Cells must be able to differentiate into desired phenotypes.
- Cells must adopt appropriate three-dimensional structural support/ scaffold and produce ECM.
- Produced cells must be structurally and mecha-nically compliant with the native cell.
- Cells must successfully be able to integrate with native cells and overcome the risk of immun-ological rejection.
- There should be minimal associated biological risks.

The source of cells utilized in tissue engineering can be autologous (from the patient), allogenic (from a human donor but not immunologically identical), or xenogenic (from a different species donor).

Autologous cells represent an excellent source for use in tissue engineering because of the low association with immune complications. Autologous cells are however not cost effective and batch controlled for universal clinical use. In contrast, allogenic cells offer advantages over autologous cells in terms of uniformity, standardization of procedure, quality control and cost effectiveness.

Cell sources can be further delineated into mature (non-stem) cells, adult stem cells or somatic stem cells, embryonic stem cells (ESCs), and totipotent stem cells or zygotes.

The utility of mature cells is restricted because of its low proliferative and differentiating potential. Adult stem cells are resident stem cells found in specific niches or tissue compartments and are important in maintaining the integrity of tissues like skin, bone and blood.

They are undifferentiated cells that can be programmed to differentiate into specific tissue types. At least 20 major categories of somatic stem cells have been identified in mammals.

They can be found in bone marrow, blood, cornea and retina of the eye, dental pulp, liver, skin, GI tract, and pancreas. ESCs are derived from the inner cell mass of the pre-implantation blastocyst. They are undifferentiated, immature cells that are capable of unlimited self-renewal with the ability to differentiate.

Stem Cells (3 types)

Rapid advances are being made in stem cell research with a focus on their therapeutic potential for regenerative medicine and other biomedical applications. In combination with tissue engineering, stem cells hold a number of promises in further advancing contemporary medicine.

Traditionally, adult stem cells were believed to form a small number of cells restricted to a particular germ layer origin; however, some evidence now indicates that adult stem cells isolated from various tissues have greater plasticity than previously thought.

Several researchers have attributed this apparent plasticity of adult stem cells to developmental signals-mediated differentiation. For example, Azizi et al., demonstrated that marrow stromal stem cells transplanted into the rat brain can acquire a neural phenotype.

Neural and muscle-derived stem cells isolated from cloned mice were able to differentiate into all cell types of hematopoietic lineage. A year later Clark et al., reported that murine neural stem cells could differentiate into all three germ layers with appropriate culture modifications.

In 2000 Lagasse et al., demonstrated that the injection of murine bone marrow stem cells into a mutant mouse line with progressive liver failure were able to form hepatocytes and restore liver function.

Recently, however, others reported cell fusion, rather than signal-mediated differentiation, as a possible mechanism for transdifferentiation.

For example, Alvarez- Dolado et al., demonstrated the evidence for cell fusion of bone marrow-derived cells with neurons and cardiomyocytes and the lack of trans-differentiation without cell fusion in these tissues, suggesting that cell fusion may be necessary for the observed plasticity of adult stem cells.

This finding is not without controversies. Later, others demonstrated that stem-cell plasticity is a true characteristic of neural stem cells, suggesting that transdifferentiation can be accomplished without cell fusion. In summary, the concept of adult stem cell plasticity is new, and the phenomenon is not thoroughly understood. The ESC is defined by its origin, that of a blastocyst before uterine implantation.

Three unique characteristics define primate ESCs:

- They are derived from the four to five day-old embryos that have been produced in an in vitro fertilization clinic.
- They possess a prolonged undifferentiated quality that is capable of self-renewal.
- They demonstrate stable developmental potential to form the derivatives of all three embryonic germ layers ('pluripotent') even after prolonged culture.

Pluripotent stem cells have been derived in mice and primates from pre-implantation embryos and bone marrow stroma. They can be induced to differentiate into all cell types and are able to colonize tissues of interest after transplantation. Human ESCs are mostly obtained from discarded embryos generated in fertility clinics.

In the presence of leukemia inhibitory factor (LIF) or embryonic fibroblast feeder layers, ESCs can be maintained and expanded in an undifferentiated, pluripotent state in vitro almost indefinitely. Upon withdrawal of LIF, ESCs spontaneously differentiate to form distinct embroid bodies, which contain differentiating cells of ectodermal, endodermal and mesodermal lineage. Much of the development in manipulating ESCs have been based on mouse ESC models, and there are numerous examples of improved tissue function with this usage.

Human ESCs, like no other ESCs, differentiate into somatic cell types that make up the human body.

The potential benefits to health care are enormous, ranging from generating neurons for treating patients to Parkinson's disease to learning about the molecular processes that drive tumor development.

Current Progress

The current utilization of stem cells in tissue includes skin, blood vessels, cartilage, heart tissue, liver, pancreas and neural tissue. Skin defects are primarily treated by the use of epidermal and dermal constructs. Dermal fibroblasts are obtained from neonatal foreskin, expanded in vitro, seeded onto a scaffold of polylactic or polyglycolic acid before being cultured in a bioreactor system to generate a dermal layer. A bilaminate construct is produced by coating the dermal layer with multiple layers of keratinocytes. The long-term viability of these skin grafts depends on the population of the engrafted skin stem cells. Various techniques have been utilized to promote neovascularization. More recently, cell-based therapies known as endothelial progenitor cells or angioblasts have been utilized.

The lack of vascularity renders cartilage less susceptible to ischemic insult than tissues with greater dependency on direct vascularity. Autologous cell products have been used effectively for cartilage repair. Alternatively, both adult stem cells and ESCs are being investigated for their potential to differentiate into the appropriate cells for repair of damaged bone, cartilage, and tendon.

The complexity and specialized conducting infrastructure of the heart have posed a challenge to duplicate in engineering systems. Furthermore, such effort is further limited by a low proliferative potential of cardiomyocytes. Available engineered heart products are biocompatible, non-biodegradable mechanical. These devices are ineffective for longterm replacement.

Nonetheless, the possibility of development of an engineered heart is exemplified by the successful manufacturing of tissue-engineered valves and myocardial infarct scar remodeling. Combining the present engineering efforts with the potential clinical applications of stem cells may accelerate the rate of the development of these engineered products. Kehat et al., first described the generation of a reproducible spontaneous cardiomyocyte differentiating system from human ESCs.

Injection of bone marrow cells in the contracting cardiac wall bordering myocardial infarcts has been demonstrated to regenerate myocardium. The development of human ESC lines and their ability to differentiate to cardiomyocytes holds great promise.

Nevertheless, bridging the gap between the theoretical potential of this unique system and any actual clinical applications will require much work. Presently, liver transplantation is the only successful treatment modality for end-stage liver failure. The limited availability of donor livers has turned our attention to alternative sources including tissue engineering.

A successful construct employs the use of highly porous biodegradable discs to deliver hepatocytes. This approach is limited by the low proliferative potential of hepatocytes and the suboptimal cell survival from the poor access to vascularity for nutrient provisions and removal of metabolic waste. Several types of stem cells, such as human adult hematopoietic and mesenchymal stem cells, have been demonstrated to differentiate into hepatocytes under the appropriate environment, and hence may serve as a unlimited cell source for therap.

Diabetes may be potentially treated by generating insulin-producing cells from stem cells and grafting them into the pancreas in diabetic patients. A much better understanding is necessary for the mechanisms that govern the expansion and differentiation into beta cells.

Clinical islet cell transplantation for the treatment of type 1 diabetes has been introduced with enthusiasm.

The procedure is considered minimally invasive and potentially offers the possibility of being performed under donor-specific tolerant conditions. Recently, multipotent precursor cells from the adult mouse pancreas have been identified, and may be a promising candidate for cellbased therapeutic strategies.

Unlike isolated beta cells which release insulin in a monophasic, all-or-none manner without any fine-tuning for intermediate concentrations of blood glucose, pancreatic precursor cells may be cultured to produce all the cells of the islet cluster in order to generate a population of cells that will be able to coordinate the release of the appropriate amount of insulin to the physiologically relevant concentrations of glucose in the blood.

The potential of ESCs in the treatment of brain diseases is also under active investigation. Ma et al., recently demonstrated the first CNS stem cell-derived functional synapse and neuronal network formation on a three-dimensional collagen gel.

Mesenchymal stem cells are ideal for replacing the dead neural cells due to their strong proliferative capacity, easy acquisition, and considerable tolerance of genetic modifications.

Following the transduction of brain-derived neurotrophic factor (BDNF) via recombinant retroviral vectors into the human mesenchymal stem cells, nearly all cells were demonstrated to express BDNF, and the quantity of BDNF in the culture medium was increased by approximately 20,000-fold. Another area under investigation is genetic manipulation of adult stem cells as a potential therapeutic strategy to surmount the problem of organ shortage.

Oncology studies have identified numerous genes involved in controlling cell growth. By introducing growth-promoting genes, many stem cells can be immortalized and gain the ability to differentiate.

For example, immortalized human hepatocyte stem cells grafted into rats with acute liver failure can keep the animals alive. There is however some concern regarding oncogenic risks with using growth-promoting genes. This problem may be unlikely by introducing three levels of safeguards:

- Deletion of the transferred oncogene by site-specific recombination followed by different selection,

- Allogenic transplantation requiring temporary immunosuppression,
- Introduction of a "cellsuicide" gene that can be activated by administering a drug.

While the selection of appropriate cell source remains a key to the successful creation of stable, complex three-dimensional constructs, equally important is the design of robust scaffolds and ECM that mimics a native physiologic environment and a capillary network in a three-dimensional structure. In collaboration with Draper Laboratories, Shieh et al., have developed a computational model of vascular circulation.

Future of Tissue Engineering with Stem cell Research

Although much progress has been made in the field of tissue engineering, further work toward organ and tissue replacement is necessary. The optimal cell source, scaffold design, and in vitro bioreactors, the use and development of microfabrication technology to create vascularized tissues and organs are still being investigated.

The search for and use of an appropriate multipotent or pluripotent stem cell in tissue engineering is an emerging concept.

Certainly, many areas of stem cell research and their potential clinical applications are associated with controversies; therefore, it is important to address the ethical, legal, and social issues early. Many technical questions are yet to be answered and require close interdisciplinary collaborations of surgeons, engineers, chemists, and biologists, with the ultimate goal of functional tissue restoration.

As more scientific knowledge will be gained from stem cell research, hopefully, some of the current ethical and technical concerns will be answered or removed in the future.

Proposition Passage in California

The California Stem Cell Research and Cures Initiative, recently passed by statewide ballot in November 2004. As a result, the California Institute for Regenerative Medicine (CIRM or Institute) was created in order to provide breakthrough cures for devastating diseases. These diseases include, but are not limited to, diabetes, heart diseases, cancer, osteoporosis, multiple sclerosis, Alzheimer's disease, Parkinson's disease, Lou Gerig's disease and spinal cord injuries by utilizing tax-free state bonds to fund stem cell research.

The institute would ensure that strict ethics codes will be adopted and enforced, including utilizing guidelines developed by the National Institutes of Health. Mirroring the existing state law, the institute would strictly prohibit human reproductive cloning.

In addition to finding cures for various diseases, the goal of the institute is also to reduce the rapidly rising healthcare costs in California and boost its economy. Proposition is designed not to burden taxpayers, authorizing tax-free state bonds that will provide an average of $295 million per year over ten years to support stem cell research in California. It assumes that a large share of Health care costs is caused by diseases and injuries that can potentially be cured with stem cell therapies.

Opponents to Proposition have argued that the bill places an unnecessary and unwise financial burden on an already-insufficient state budget. According to some estimates, state bond debt, which amounted to $33 billion in May of 2004, might rise to more than $50 billion by July of 2005, largely due to the costs of Proposition.

In addition, the initiative has yet to assure the Californian taxpayers whether the state will actually receive any royalties from discoveries paid for by state dollars. Questions about significant conflicts of interest on the Independent Citizen's Oversight Committee, (the "board of directors" of CIRM) and even about the scientific integrity of the stem-cell lines available for research have been raised.

In spite of this opposition, many defend the economic merits of Proposition. As no payments would be demanded within the first five years of the bill's implementation, the state would remain free from financial stress and allowed time to balance its budget. No tax increases would accompany the legislation.

The potential effects that the passage of California's Proposition would have on other states and the nation as a whole are not to be underestimated. New Jersey, the nation's first state to create a publicly funded institute for stem cell research and the second state to pass legislation legalizing the research, has found it necessary to review its policy and is considering an increase in their research grants.

Other states, including Massachusetts, Maryland, Wisconsin and Connecticut, are also jumping in with proposals to fund research cloning.

These changes are happening against the recent passage of the "declaration" of banning all forms of human cloning by the United Nations. A recent Swiss law that mirrors laws in most European countries, allows medical research on stem cells taken from human embryos but bars cloning. Britain and Belgium allow embryo creation for research. Britain has an independent embryo authority to scrutinize and monitor all projects and has issued only two cloning licenses since 2001.

What has made the United States such a fertile ground for expanding embryo research is not its liberal laws but the lack of them. Congress has tried but failed to pass legislation largely because of irreconcilable differences over when life begins.

Both Presidents Bill Clinton and George W. Bush banned the National Institutes of Health on ethical grounds from funding the creation of human research embryos, although their orders apply only to federally funded work.

In 2001, Bush additionally narrowed NIH research to embryonic stem cells already harvested from spare IVF embryos. These restrictions in turn triggered an influx of private and state funding, and the dramatic Proposition.

Limitations

One of the major challenges of the field is to design an appropriate capillary network to allow gas exchange, provide nutrition, and remove metabolic waste from the implanted constructs. It will be necessary to create a tubular system with a minimum diameter of 10 ìm to allow vascular ingrowth, especially important for multilayered solid organ constructs with a volume greater than 2 to 3 cubic mm. An alternative is to implant a vascularized construct in an ectopic, well-vascularized location, such as omentum.

Special attention will be given to the control of the local milieu of each cell type as different cell types often require different culturing environments, making it difficult to design a multilayered organ system construct. A challenge will be to create a scaffold capable of supporting various cell types. The appropriate expression sequence, dosing, and duration for growth factors also needs to be defined if we are to maximize ESCs utility in tissue growth and maintenance in the pluripotent state.

Ethical/technical Considerations

There are three main research programmes in stem cell research.

- Research on adult stem cells.
- Research on ESCs from discarded embryos that are produced via in vitro fertilization.
- Research on ESCs obtained via therapeutic cloning.

There is little controversy surrounding human adult stem cells. On the other hand, human ESCs face great controversies, the extent of which is dependent on the type of research programmes.

Various autologous adult stem cells, including mesenchymal, hematopoietic, neural, muscle, and hepatic stem cells, are actively being investigated. They are more likely immunocompatible and are not associated with ethical concerns unlike their ESC counterparts. Challenges remain in regard to optimization of isolation techniques that minimize contamination, permanently maintain the desired cell types following differentiation, and increase the production of a large number of cells that are adequate for the construction of a tissue or organ.

Furthermore, much of the data cited have not been reproducible. The utility of adult stem cells would effectively avoid the ethical problem of using ESCs. But much needs to be learned about adult stem cells in regard to their replication and differentiation. ESCs may offer technical advantages over other sources of somatic precursors. For example, ESCs can be generated in abundant quantities in the laboratory and can be grown in their undifferentiated state for many generations.

In contrast, researchers have had difficulty finding laboratory conditions under which some adult stem cells, especially hematopoetic stem cells from blood or bone marrow, can proliferate without becoming specialized. This technical barrier to proliferation has limited the ability of researchers to explore the capacity of certain types of adult stem cells to generate sufficient numbers of specialized cells for transplantation purposes.

Unlike adult stem cells, ESCs can be grown indefinitely in tissue culture. There is an increased risk for immune rejection because of genetic mismatch, which in turn would demand lifelong high-dose immunosuppressant treatment. Furthermore, the creation of human ESCs derived from discarded, non-transferred human embryos using an immunosurgical technique has resumed arguments that have not been resolved.

A SART-RAND study identified at least 400,000 frozen embryos in storage since the late 1970s. Only 2.8 per cent of these embryos have been designed for research. With a 15 per cent efficiency rate for generating a human ESC line from blastocysts and incorporating some losses to the freeze-thaw process, only approximately 275 human ESC lines can be created from the excess. The pluripotential virtue of stem cells itself renders limitations: the efficiency of differentiation appears to be compromised for certain cell types, such as hepatocytes and cardiomyocytes, and the quality of differentiation is still questionable.

The use of ESCs as an important cell source for tissue engineering has great potential for the treatment of diseases, but there are alternatives, such as the use of adult stem cells and ESCs from therapeutic cloning under active investigation. Worldwide legislation reflects the dilemma, ranging from the more permissive to the nominally more restrictive (the United States—only minimal federal financial funds granted to research that meets the strict NIH guidelines, but stem cell research is legal without any restriction in the private sector) to the truly restrictive (Germany, Austria, Ireland, France—laws on reproductive medicine ban the extraction from stem cells from a human embryo).

The host immune response to allogenic embryonic stem cells poses a major challenge in addition to ethical and religious concerns regarding ESCs. This has led researchers to investigate creation of a universal donor cell by making histocompatible proteins on the cell surface thus reducing the cells' antigenicity.

This is the heart of nuclear transfer technology or "therapeutic cloning". However, the technology has its own shortcomings. It has a low efficiency rate which in turn requires a large number of oocytes. Its inability to control differentiation requires a better understanding of ESC growth, differentiation, and application in engineering tissues. Furthermore, therapeutic cloning protocols in vitro culture systems must be optimized before contemplating the use of this technology for cell therapy.

All three research programmes discussed here are directed at increasing our knowledge about basic cell biology, creating new therapies through stem cell culture and control of cell differentiation, and producing commercially viable stem cell products either by the direct patenting of stem cell lines or by combining stem cell technologies with genetic engineering or other patentable interventions.

The primary discussion on stem cells is concerned with the ethical issues raised by each of these programmes and with whether these issues should influence regulatory decisions regarding public financial support for research.

3

Industrial Fermentation

'White' biotechnology or industrial and environmental biotechnology is a broad and expanding field that includes making enzymes with a variety of industrial uses that include the manufacture of bioplastics and biofuels and using micro-organisms and plants for the treatment of wastes and abatement of pollution, a process known as bioremediation.

It may seem to be occupying a small niche now, but its products or processes will become as ubiquitous as those of the chemical industry today because of the benefits they pose to the environment and the premium attached to this environment-friendliness for the companies that undertake them. Bacterial bioleaching, i.e. using specific bacteria to extract metals from ores or mine wastes sprayed with water, is a growing sector of the mining industry and several developing countries are already playing a key role in this area.

Other research efforts are scanning the microbial diversity of various environments (e.g. the oceans and seas) and deciphering their genetic information aimed at isolating micro-organisms that could be used in the manufacture of drugs, enzymes and a wide range of bioactive compounds, as well as in bioremediation processes.

As for materials and fuels that are not derived from petrochemical processes, researches are also underway for processes to improve and enhance the bioremediation of water, soils and ecosystems at large, and minimize the use of fossil-fuel energy. All these forms of 'white' biotechnology are poised to increase the field's positive social acceptance.

DEFINITION AND POLICIES

The so-called 'white' biotechnology is a broad and expanding field that makes use of new enzymes for a variety of industrial uses, embraces the manufacture of non-oil-based and biodegradable bioplastics and biofuels, as well as artificial fibres and encompasses many treatments of wastes and abatement of pollution, using micro-organisms and plants, known as bioremediation. White biotechnology is part of the contribution of applied science to a healthier environment and to sustainable development. 'White' biotechnology or industrial and environmental biotechnology may now only seem to occupy a small niche but because of its enormous promise and potential, it may become as ubiquitous as those of the chemical industry today.

Some analysts have compared the current status of biotechnology to that of chemistry in the 1870s, when chemists were applying their new knowledge to a limited range of applications (e.g. dyes). Now, industrial chemistry is everywhere, from agrifood and pharmaceutical industries to fuel production to textile, fertilizer, paper industries, etc.

From the policy viewpoint, after the pioneering efforts by the US National Academy of Sciences at the end of the 1980s, the future-oriented studies initiated in 1996 and the first report by the Organisation for Economic Cooperation and Development (OECD) in 1999 on 'white' biotechnology, a decree signed by the US President William Clinton on 12 August 1999 on 'Developing and promoting bio-based products and bio-energy' stated that between 2000 and 2010 the consumption of biotechnology-derived products and fuels should be trebled.

In June 2000, the Biomass Research and Development Act and the Sustainable Fuels and Chemicals Act, voted by the US Congress, allocated hundreds of million dollars to research projects jointly carried out by universities and industry; they also created the Biomass R&D Board, in charge of coordinating the action of federal administrations (commerce, energy, agriculture, etc.).

Later on, US President George W. Bush strengthened this framework in the Farm Bill. Behind this decision was the need to find outlets for agricultural surpluses, to ensure national security by decreasing the dependence on fossil fuels, and to poise US farmers to become the future suppliers of feedstock to the industry. The US industry could also look at this as a way to boost their competitive edge over European corporations.

Meanwhile, in Europe, the Biowise Programme of the British government had surveyed 70,000 industrial installations, which could reduce costs and decrease pollution through biotechnology. The government wanted to convince these industries to invest in this area.

BASF AG in Germany, the Dutch chemicals company DSM and most of Scandinavian paper-manufacturers were already doing so.

The European Union's heads of state and government, at the 2001 Göteborg Summit, stressed the need for sustainable development in Europe, and in March 2002 in Barcelona, they underlined the key role of biotechnology to achieve it.

On 14 July 2002, the European Commissioner for Research announced that □3.42 billion were to be devoted to the programme 'Clean technologies' between 2002 and 2006. Earlier on, on 4 July 2002, the European Parliament voted a directive on biofuels that indicated a target of 5.75 per cent for the proportion of biofuels consumed for transport by 2010, a great increase from the 0.3 per cent in 2002.

In its long-awaited environmental plan (ETAP), released at the end of January 2004, the European Commission identified 'white' biotechnology as an "important technology" with a bright potential for environmental benefits. The action plan foresaw opening up new financial resources from the LIFE Programme or structural funds to finance pilot and demonstration projects.

Additionally the Commission aimed to put in place some solid science to improve the market acceptance of the innovation technology so they also envisioned the creation and funding of expert networks that should validate the biotechnological processes.

'ETAP was designed to deliver environmental benefits and give a boost to innovation', stated Steen Riisgaard, chair of the Industrial Biotechnology Board at the European Bio-industries Association.

ENZYMES

Purifying and selling bacterial enzymes for use in cheese making, food manufacturing, cotton weaving, washing powders, etc., has existed for decades, but in 1988 the Danish company Novozymes produced the first transgenic enzyme, a fat-digester for detergents. The company supplies, among others, enzymes for the scouring process, which consists of removing the brown, non-cellulose parts of cotton that in the conventional process involves the use of relatively harsh chemical solutions.

Thanks to advanced genetic techniques, multiple enzyme variants can be created at high speed, which are then screened to fit the desired applications. Novozymes has become the world's biggest enzyme manufacturer, and also renowned for its transparency procedures, for involving the consumers in the discussion of its activity report, and its commitment to a healthier environment and to sustainable development.

Genencor, created in 1982 as a partnership between the biotechnology company Genentech, Inc., and the chemical firm Corning, had an annual turnover of $325 million in 2001 and a portfolio of 3,400 patents and licenses. Its main income is from the sales of enzymes, which are discovered, tested and standardized in Genencor's laboratories at Palo Alto (California) and Leiden (Netherlands).

At the Leiden laboratory, established in 1995 in the BioScience Park, some 25 researchers are trying to improve the performance of enzymes developed for Procter & Gamble, Unilever and other detergent manufacturers (by 2002, the detergent market value was estimated at $400-$500 million a year). Genencor is famous for having isolated an enzyme that gives the stone-wash colour to jeans, thereby replacing the process, which consisted of washing the jeans in machines with volcanic stones.

Genencor's enzymes are being massively produced in bioreactors located in the USA, United Kingdom, Belgium, Finland and China. In addition to the enzymes mixed with detergents, the company had signed a $17-million contract over three years to work on enzymes involved in the production of biofuels. Genencor is also involved in vitamin-C production with Eastman Chemical, biopolyester manufacture with E.I. DuPont de Nemours & Co., Inc. (DuPont), and sugar with soft drink companies.

Genencor is also trying to develop enzymes for manufacturing hypoallergenic proteins, which would penetrate the cosmetics and body-hygiene industry – a $1 billion market. The company is interested in developing biosilicones in association with Dow Corning which brought in $35 million. Genencor is looking to two to three years to extend the market it would have captured and create new ones in four to ten years. Meanwhile, it will patent technological innovations while waiting for a market that may emerge in the longer term.

Sigma-Aldrich Fine Chemicals, a division of Sigma-Aldrich Corporation, well known for its expertise in protein production and processing, agreed with ProdiGene, Inc., a Texas-based biotechnology company, to manufacture and distribute TrypZean, an animal-free recombinant trypsin, produced with ProdiGene, Inc.'s proprietary transgenic plant system. Trypsin is an enzyme normally produced in the pancreas of animals that breaks down proteins as part of the digestion process.

It is also a critical enzyme in the bioprocessing of proteins, such as insulin, where it is used to help cleave the protein into its active form. Trypsin is widely used in cell-culture applications for research on, and production of, recombinant proteins for clinical uses.

Another application is its use in the wound-care markets as an oral treatment of inflammatory oedema, haematoma and pain associated with a wide variety of internal and external wounds. The enzyme also has uses in food processing and in the making of infant formulas. Concerns about the use of animal- and human-derived proteins in the manufacture of biologicals make ProdiGene, Inc.'s expertise in the area of plant-produced recombinant proteins an attractive alternative. The proprietary technology offers an animal-free source of proteins as well as the capacity to produce large volumes that can be easily scaled-up or down.

Extremophiles, i.e. micro-organisms living in very harsh natural conditions, have attracted the attention of scientists who are collecting them from volcanic areas, hot springs, deep-sea vents, soda lakes, arctic rocks and industrial effluent outlets.

This practice called "bioprospecting" is followed by the study of metabolism of the micro-organisms and the isolation of enzymes and other proteins with extraordinary properties, such as tolerance for heat, cold or salt. These properties make them suitable for use in the manufacturing of washing powders, in textile industry and bioremediation processes.

Japanese companies are very active in this bioprospecting and research area, as well as Costa Rica whose great biological diversity is not only due to the preservation of its rainforests but also to the existence of numerous extreme environments (mainly volcanic areas). Other firms like Applied Molecular Evolution, Genencor and Maxygen, interested in the search for drugs, are attracted by the extremophiles, and their peculiar metabolism and evolution.

India too is a player in biotechnology with the bulk of India's industrial biotechnology sector focuses on producing enzymes for the textile and paper industries. The sector was the second-largest exporter ($37 million in 2002-2003) to other Asian and European countries. In 2002-2003, this sector produced revenues of $72 million and was rapidly growing.

The example of vitamin-B2 synthesis illustrates how a new biotechnology process can benefit both environment and economics. This vitamin is generally produced using a complex eight-step chemical process. BASF AG's new process reduces it to a one-step process involving fermentation whereby the raw material is fed to a mould, which transforms it into the finished product, recovered as yellow crystals directly from the fermentation broth.

The biotechnological process reduces overall costs by up to 40 per cent and the environmental impact by 40 per cent (CO_2 emissions are reduced by 30 per cent, resource consumption by 60 per cent and waste by 95 per cent). The synthetic pathway to the antibiotic cephalexin has also evolved from a multistep chemical process to a mild bio-transformation, based on a fermented intermediate linked enzymatically with a side chain to the final end product. The biotechnological process uses less energy and input chemicals, is water-based and generates less waste.

BIOPLASTICS

Henry Ford, a great American inventor, was fascinated with soybeans. During the Depression, he spent many hours in a laboratory near Detroit trying to transform the beans into cheap plastic, so that more Americans could afford his Model T car. By 1941, Ford eventually unveiled a hand-made car with a plastic body made completely from plants but unfortunately, he never figured out how to produce soybean-based plastic that could compete on price with oil-derived plastics, still the dominant type of industrial polymers.

Currently, the two most advanced plastics projects are those of E.I. DuPont de Nemours & Co., Inc., one of the world's biggest chemical corporations (which invented nylon), and of Cargill-Dow Chemical Company, a joint venture between an agricultural and a chemical firm.

Established in 1865 and headquartered in Minnetonka, Minnesota, Cargill, Inc., described itself in its 1997 financial report as 'an international marketer, processor and distributor of agricultural, food and industrial products with some 79,000 employees in more than 1,000 locations in 72 countries and with business activities in 100 more'. Cargill, Inc.'s global revenues peaked at $56 billion in 1996 and 1997, dropping somewhat to $49.4 billion in 2001. Dow Chemical Company is headquartered in Midland, Michigan.

DuPont's process, developed in collaboration with Genencor, consisted of assembling in a single bacterium biochemical pathways from three different micro-organisms. The raw material for the process is glucose syrup made from maize starch, which is converted into 1,3-propandiol, used to manufacture a polyester called Sorona. The latter is not entirely a biopolymer, but a copolymer, made of 1, 3-propandiol and terephthalate, which still has to be made from oil. Cargill-Dow Chemical Company's product, Ingeo, is made out of lactic acid, which in turn is produced from glucose derived from non-transgenic maize starch.

Polymerization of lactic acid, through traditional techniques leads to poly-lactic acid (PLA), i.e. Ingeo or NatureWorks PLA. In January 2002, Cargill-Dow Chemical Company opened a $300-million plant at Blair (Nebraska) that could manufacture 140,000 tons of Ingeo, biomaterials to be mainly used in packaging and bedding.

This plant was the first to produce biomaterials on a commercial scale. As Ingeo was more expensive than petrochemical plastics, Cargill-Dow Chemical Company intended to introduce it in the market of environment-friendly goods, which were sold at a higher price.

This friendliness is due to the fact that no (or little) fossil fuels are used to manufacture the biopolymers and that they are biodegradable so they do not contribute to global warming or to pollution. Coca-Cola Co. is now using Ingeo to make soft-drink cups, McDonalds for salad containers, and Pacific Coast Feather Co. to fill pillows and comforters. If the production cost of bioplastics could be halved, the amount produced in 2010 could be trebled.

Small firms are also interested in the manufacture of biopolymers. Metabolix, a company based in Cambridge, Massachusetts, is working on the production of monomers by living organisms and on their polymerization by the same organisms. James Barber and Oliver Peoples, the founders of Metabolix, working in the early 1990s at the nearby Whitehead Institute, realized that some bacteria could store polyhydroxyalkanoate up to 80 per cent of their weight.

Polyhydroxyalkanoates (PHAs) are produced by either non-recombinant bacteria (e.g. Ralstonia eutropha) or recombinant ones (e.g. Escherichia coli to which genes of R. eutropha have been transferred). The production cost of bacterially produced PHA was between $3 and $5 per kg, while that of propylene was less than $1 per kg. There is therefore a need to decrease the production cost and also increase productivity. Metabolix planned to start commercial production of PHA by the end of 2003, at a price comparable with that of the more expensive polyesters.

The company hoped that biodegradability and tougher regulation on the disposal of synthetic plastics into the environment would give a competitive edge to its products over the tried and trusted chemical materials.

In 1989, researchers of the French Development Research Institute (IRD) discovered in palm wine the bacterium Bacillus thermoamylovorans, which is able to produce lactic acid from sugar-rich wastes. In cooperation with a subsidiary of France's third-biggest sugar producer (Erstein), Epicure, the first production trials were carried out in April 2000 on three kinds of juices which demonstrated the efficiency of the process and led to high yields of lactic acid (100g/litre). The bacterium grows at a temperature between 47°C and 58°C – which avoids the contamination by other micro-organisms.

The Research Centre for Artificial Biopolymers (CRBA), in Montpellier, south-east of France, has also been developing a polymerization process of lactic acid, in order to produce bioplastics on a commercial scale.

All companies working in this field are trying to bring down costs. Metabolix, for instance, hopes to switch from producing plastics from bacteria (which need to be fed) to manufacturing them in plants (through photosynthesis). The process is being scaled up from successful laboratory trials. It is also foreseen to produce plastics in transgenic plant species.

In 1992, French researchers were able to transfer two genes of R. eutropha into Arabidopsis, which could produce 0.1 per cent of its biomass as PHA. In 1994, this proportion was increased up to 14 per cent after the transfer of a third gene and a better expression of alien genes.

This yield was considered satisfactory, but the polymer had poor physical properties. In 1998, small quantities of a good quality PHA, called MCL-PHA, could be derived from a transgenic plant; this plastic resembled elastomeres, rubber or glue and since the market for elastomere-like materials was less important than that of plastics used for packaging, large-scale production was considered less crucial.

Clint Chapple a biochemist of Purdue University, and Knut Meyer of E.I. Dupont de Nemours and Co., Inc., have cloned a gene of Arabidopsis which could influence the synthesis of plastics in plants without harming their structure. The scientists stated they would be able to produce plastics that could serve for the manufacture of heart valvules, in the building of aeroplanes and other sophisticated uses.

Plants could produce the monomeric units of the plastic polymers, as some monomers are difficult to elaborate from oil. C. Chapple's work consisted of trying to stabilize some monomers and to produce them in high concentrations in plants. The viability of the whole process depended on the storage of sufficient amounts of the plant-made plastics in the cell vacuoles from which they could be extracted. The gene was eventually patented in favour of the University and the US corporation.

The results published by Steven Slater and his collaborators in the October 1999 issue of Nature Biotechnology also opened new prospects for the large-scale production of a very good quality PHA from the leaves of transgenic Arabidopsis as well as from the seeds of transgenic oilseed rape. The US researchers had transferred four bacterial genes to the plants that could be expressed in two distinct metabolic pathways.

However, both transgenic Arabidopsis and oilseed rape produced only 2.5 per cent of their biomass as plastic, which implied that more genetic transformations would be necessary to obtain commercially exploitable results. The extraction process was also an important element of a profitable production of plastics by transgenic crops.

There is also a research consortium comprised of Malaysian scientists and scientists of the Massachusetts Institute of Technology that has been working on the development of transgenic oil-palms that produced a bioplastic in their leaves. The genes transferred to the tree were those of Ralstonia eutropha, which control the synthesis of polyhydroxyalkanoate and polyhydroxybutyrate – the precursors of the bioplastic.

The latter was to be recovered from dried leaves, currently used as feed or for fibre production; it will not be present in the fruit bunches, used to extract palm oil. By the end of 2003, trials were expected to be carried out to test the production capacity of transgenic plants.

FEEDSTOCKS AND BIOFUELS

Another promising biotech idea is to use the whole plant as a chemical feedstock. Glucose syrup is a refined product, made out of maize starch, i.e. maize grains. The latter costs about $80 a ton; that is cheaper than oil, weight for weight, but this could be improved if the plant waste is also used, as it fetches only about $30 a ton for silage.

But it consists mainly of cellulose, also a polymer of glucose, but more difficult to break down; therefore, the search began for efficient cellulases. Verdia, Maxygen's plant-biotechnology subsidiary, is trying to develop a cellulase that the plant would make in its own cell walls, while preventing the enzyme to digest the living plant. If research and development were successful, limitless supply of glucose from the degradation of cellulose would follow, as well as all kinds of bioplastics.

Ethanol Production

The OECD (Organisation for Economic Cooperation and Development) and a handful of big corporations already acknowledge that the fossil-fuel era will come to an end and that industry has to prepare for it. Governments are helping drive the shift to alternatives through subsidies and regulations. The USA, for instance, published draft rules to encourage federal purchasing of bio-based industrial products in 11 categories, from lubricants to fibres, plastics and paints. It also subsidizes making ethanol from maize kernels.

The EU targets under the Kyoto Protocol, have their eye on some 9.3 million tons of ethanol that should be produced annually in Europe by 2010. The biofuels target for 2005 is 2 per cent of vehicle fuel, rising to 5.75 per cent by 2010.

Biofuel production is part of 'white' biotechnology. Ethanol, a biofuel, is produced from the fermentation of cane sugar. In Brazil, they have mastered the technology as well as that of building motor engines that use this rather corrosive fuel.

The decision was made to decrease the amount of the oil bill and to seek energy independence. Ethanol served to power the first car of Henry Ford, while nowadays in Brazil an important proportion of the car fleet uses a mixture of gasoline and 20 per cent ethanol. Even in the USA, nearly a tenth of all motor fuel sold is a blend of 90 per cent petrol and 10 per cent ethanol.

The American market for bioethanol is 8 billion litres a year. At Genencor, some reckon it could be as high as 75 billion litres a year by 2020. That would be enough to replace two-thirds of US current petrol production. Royal Dutch Shell and Iogen, a Canadian firm, were among those building an ethanol refinery that converts 'residual biomass' – leftover plant matter usually thought of as waste. Thanks to the creation of more efficient enzymes that cut fermentation costs, advances in transgenic maize that have improved yields, and government subsidies, 80 US ethanol refineries produced a record 11 billion litres in 2004, or less than 2 per cent of all motor fuel used.

This figure is already very close to the US Congress target of 19 billion litres of maize-derived ethanol (2.5 per cent of all motor fuel used) set by year 2012.

Canada intended to quadruplicate its ethanol production, up to 1 billion litres, between 2000 and 2005. Brazil – the world's first-biggest producer of ethanol –produced 15 billion litres annually.

In 2003, Europe produced 700,000 tons of biofuels, half of it in France. While the production, transport and consumption of gasoline generate 11.8 kg of CO_2 per gallon (3.8 litres), in the case of ethanol 7 kg to 10 kg of CO_2 are generated if conventional production processes are used, but only 0.06 kg if one relies on bioprocesses.

Although motor fuels mixed with ethanol could cut hydrocarbon and carbon monoxide emissions, they also produced aldehydes and peroxyacyl nitrates, which could form brown smog. Ethanol producers have previously fallen foul of the US Environmental Protection Agency (EPA) for emitting volatile organic compounds such as formaldehyde and acetic acid, and methanol.

Techniques for efficiently disposing of large volumes of these chemicals must be developed and strictly applied.

Biofuel Production in the European Union

In the European Union, in 2003, two directives have set the following targets for biofuel content in the currently used fuels: 2 per cent in 2005 and 5.75 per cent in 2010. The biofuels produced in the European Union's member countries, particularly in France are the following:

- Methylester of vegetable oil or biodiesel (diester), produced from groundnut, oilseed rape or sunflower seeds; soybeans and palm oil can also be used; the extracted oil is transformed into methylester, incorporated to gasoil by the oil-industries;
- Ethanol, produced from sugar-cane, sugar-beet, wheat or maize; in Europe, it is transformed into ethertertiobutylether (ETBE) – a mixture of isobuthylene, a by-product of the oil industry, and ethanol – before being mixed with petrol; such mixture, considered technically justified by the oil industry, is questioned by ethanol-producers who recommend the direct incorporation of ethanol into petrol as it is done in Brazil and the USA.

The maximum biofuel content of current fuels is 5 per cent; it can be increased to 30 per cent in the case of collective transport means or of vehicles under public administration. In France, the average rate of incorporation of biofuels is only 1 per cent.

The volume of production of biofuels is regulated in France through the allocation of tax exemptions, which amounted to □350 per cubic meter (m3) for diester for a production quota of 387,000 tons. While 300,000 hectares were planted with oilseed rape in 2004 to produce biofuels, specialized professionals estimated that 800,000 ha could be devoted to this crop for the same purposes. With respect to ethanol, France was producing 1.3 million hectolitres on 25,000 hectares of sugar-producing crops; with the rate of tax exemption was □380 per m3·

To reach the target of 5.75 per cent of biofuel content in the currently used fuels as set by the European Commission, the following surfaces could be cultivated with wheat (309,000 ha, i.e. 7 per cent of the whole surface) and sugar-beet (62,000 ha, i.e. 14 per cent of the whole acreage), which corresponded to 3.9 per cent of the available area.

The European Commission supports the cultivation of energy-producing crops at the maximum rate of □45 per hectare.

In France, a plant built up in 1993 in the heart of the harbour and industrial zone of the city of Rouen, produces 350,000 tons of vegetable oil annually from 800,000 tons of oilseeds supplied by farms in the north-west of France.

Two-thirds of the volumes of oil thus produced were transformed into biofuels in 2004. The latter were not substitutes for current fuels, but complements that were mixed with petrol or gasoline. Such a factory bears witness to the period when France was pioneering the production of biofuels. After the reform of the Common Agricultural Policy (CAP) in 1992, which introduced the fallow scheme, farmers have been authorized to grow energy-producing crops.

Large-scale agricultural zones, such as Beauce and Picardie, seized this opportunity with the government's assistance but they soon reached a plateau in their production capacity. The forthcoming reform of the CAP in 2006 was expected to oblige farmers to find new outlets for their biofuel products.

A study carried out under the aegis of the French Agency for Environment and Mastering of Energy (ADEME) and of the Directorate of Energy Resources of the Ministry of Industry, published in September 2002, emphasized that the energy balance of biofuels was very positive: they produced more energy than they consumed in their production, in contrast with gasoil and gasoline. The development of biofuels could lead to diversifying energy supply in France, but the main obstacle to this development is their cost, which is two to three times higher than that of oil products and derivatives.

Therefore, in the absence of fiscal aid, the consumer will have to pay more for fuel. Another constraint relates to the balance between the market of petrol (gasoline) and gasoil: the European Union was importing 22 million tons of gasoil and exporting 20 million tons of gasoline, mainly to the USA.

Such a situation is explained by the strong increase in the number of diesel engines, nowadays found in one car out of three in Europe. While the demand for gasoil was increasing by 2.5 per cent annually, gasoline consumption was decreasing by 1.5 per cent.

Car manufacturers are also worried about the impact of ethanol on modern engines, compared to other mixtures like diester or biodiesel, which behaves like ordinary fuel. According to ADEME, fuel mixtures with 5.75 per cent ethanol, as recommended by the European directive, would generate problems of when it breaks down and at certain temperatures the formation of vapour bubbles that could disturb the injection systems.

However, in Brazil, motor engines work with fuel mixtures containing up to 22 per cent ethanol, but engines with injection devices are not very common in Brazil.

Biofuels could help car manufacturers meet their commitment on reducing CO_2 emissions. It was estimated that the European directive on the use of biofuels would reduce CO_2 emissions by 3.5 per cent; this reduction, although modest, was significant when the decrease in the consumption of motor-cars had reached incompressible limits, and when new devices such as hydrogen-powered cars were not expected before 2024.

While it is true that biofuels cost more than fuels derived from fossil energy, the real cost of the latter does not integrate the heavy costs of shore and sea contamination by oil and oil-tanker wreckage, nor those of conflicts involving multiple stakeholders generated by oil exploitation. Biofuels have positive effects on employment, tax recovery and supply reliability.

The impact on local employment has been evaluated at 6 to 10 jobs created for every thousand tons produced. In fact, on 28 April 2004, the French Economic and Social Council adopted a report on the 'non-food outlets of agricultural commodities' that was very favourable to biofuels.

The report stated that 'we should not wait for oil prices to reach $50 per barrel (the level that would make biofuels competitive) in order to set up substitution processes'. In addition, the rapporteur found it worrisome to see France 'being outpaced by Spain in the case of ethanol production and by Germany in the case of biodiesel', and he criticized the 'taxation disparities' in the European Union as these may hamper the development of biofuels.

The strong increase in biofuel production and the demand for their use would mean an increase in the area cultivated with energy-producing crops. Taking into account the currently available cropping areas as well as crop rotation, biofuel potential is limited to 4.5 per cent of fuel use, lower than the 5.75 per cent recommended by the European directive.

The solution may be to produce biofuels from the biomass, i.e. produce larger volumes with an energetic yield two to three times higher than ethanol. The main problem remains the magnitude of investments needed to put in place such an industrial process.

The Sun Fuel programme, funded by Volkswagen, Daimler Chrysler and Renault, has set up pilot units, but generalization of these valuable experiments would certainly need a few more years.

One way to compensate Europe's deficit in biofuels may be to import them from Brazil, especially ethanol, at lower cost. However, trade negotiations with Mercosur's member countries (Argentina, Brazil, Paraguay and Uruguay) in that respect caused panic among farmers in France, who considered that imports would kill the French endeavour to produce such biofuels.

At the same time, farmers' associations underlined that biofuels should not hinder food-crop production, considered the primary task of agriculture, and that the foreseen intensification needed to increase yields of energy-producing crops was not in conformity with the goal of a 'rational' agriculture, using less fertilizers and pesticides, and relying more on crop rotation. This opinion was shared by Greenpeace.

The Paper Industry

The paper industry's basic raw materials are trees. Separating cellulose from lignin in wood is costly and uses non-environment-friendly chemical processes. Researchers at the State University of North Carolina have bred aspens with only half the lignin of ordinary ones and, it turned out, they had the additional advantage of growing faster. However, the commercial use of transgenic trees is at least ten years off but is on its way.

The same is the case with starch, which papermakers use to bind the pulp fibres together and to 'size' the surface, so one can print on it. In Europe and North America, the starch is often extracted from potatoes. But potatoes produce two polymers making up starch: amylopectin, which papermakers like, and amylose, which they dislike. In the 1990s, the world leader in potato starch, AVEBE, a Dutch cooperative, developed a transgenic potato containing more amylopectin, less amylose, but was thwarted by the European Commission, which forbade its marketing.

AVEBE grew a new variety, though it would be years before it could reach the market. Through a Swedish subsidiary, BASF AG also created a high-amylopectin transgenic potato; the Swedish authorities gave permission for its large-scale planting in April 2004. The European Commission's consent was required for the growing of this potato variety elsewhere in Europe.

But potatoes need not be the only source of genetically engineered starch. The world was producing 190 million tons a year of cassava, nearly all for food or animal feed. But its starch too can be used for making paper, and in Thailand a little is already used for that purpose. But that could soon become a lot: Thailand is growing enough cassava to be the only significant exporter, and recently decided to allow commercial cultivation of transgenic crops.

ENVIRONMENTAL BIOTECHNOLOGY: BIOREMEDIATION

Bioremediation is a key area of 'white' biotechnology, because the elimination of a wide range of pollutants from water and soils is an absolute requirement for sustainable development. There are numerous processes of cleaning water, industrial effluents and solid wastes, using microorganisms aerobically and anaerobically. Some of them are quite sophisticated, while others are simple and adapted to the conditions of developing countries.

For instance, using microalgae (and in particular blue-green algae or cyanobacteria) in ponds to eliminate nitrogen and phosphorous, after organic matter has been degraded by bacteria, leads to water that can be recycled for irrigating non-food crops (e.g. cotton) or for industrial purposes; in addition, microalgal biomass can be used as feed.

Removal of Toxic Metals

Mercury is a highly toxic metal which, once released into water, accumulates in the food chain, damaging fish, shrimps and poisoning people who eat them. The infamous Minamata accident (called after a town on the Japanese island of Kyushu where the inhabitants suffered the toxic effects of fish poisoned by mercury-rich industrial effluents) is an example of the devastating effects of mercury on the central nervous system.

Existing techniques of mercury removal, such as precipitation or ion exchange, are expensive and not sufficiently efficient, as small but significant amounts of mercury still remain in the water.

Researchers discovered that many bacteria had developed high tolerance to heavy metals, which related to the binding of these metals to proteins, e.g. metallo-thionein that binds mercury. As naturally thriving mercury-tolerant bacteria are rare and cannot be grown easily in culture, researchers at Cornell University, Ithaca, New York, inserted the metallo-thionein gene into Escherichia coli.

A sufficiently large number of genetically engineered bacteria could thus treat mercury-polluted water inside a bioreactor. The efficiency of the procedure was high, as mercury was removed from polluted water down to a few nanograms per litre. Once the bacteria died, they were incinerated to recuperate the accumulated pure mercury.

Mercury emissions were predicted to increase by 30 per cent throughout Europe between 1990 and 2010. The European Commission funded a demonstration project to show the feasibility and profitability of the microbial remediation technology under real time conditions.

A plant was set up at Usti-nad-Labem in the Czech Republic, and has been operating since July 2000. While phycoremediation is bioremediation based on the use of micro- and macro-algae, phytoremediation, relies on higher plants to clean water and soils from heavy metals and other pollutants, or to recolonize former mining areas (e.g. in South Africa, Australia, USA, Canada, France, etc.).

For instance, heavy metals in industrial effluents can be concentrated in aquatic plants and thereafter recovered. In 2003 in Western Europe, there were about 1.4 million sites discovered to be polluted. Current remediation techniques are chemical or physico-chemical extraction techniques; they are costly and destroy soil structure. Phytoremediation techniques use “hyper-accumulating plants” which can store 10 to 500 more pollutants in their leaves and stems, which are thereafter harvested, incinerated, and metals recovered from ashes and reused in metallurgy.

The survey of hyper-accumulating plants started by the early 1990s. They are often small plants, such as Alyssum murale, which grows on metamorphic rocks, Brassica juncea, the Indian mustard, which extracts lead, or Thlaspi, which accumulates zinc and nickel. About 400 species have been identified, including 300 that accumulate only nickel.

An endemic tree in New Caledonia, Sebertia acuminata, contains up to 20 per cent of nickel in its sap and is coloured in green (nickel is generally toxic to plants at a concentration of 0.005 per cent). In the Democratic Republic of Congo, the number of plants accumulating copper and cobalt is highest: 24 and 26 species, respectively. The accumulation efficiency is not generally very high.

For some metals like silver, mercury and arsenium, there are yet no plants known to accumulate them. However, in 2000, the team of Lena Ma of the University of Florida, Gainesville, identified a fern, Pteris vittata, which tolerates and accumulates arsenium, while conserving a very rapid growth and a high biomass.

Edenspace, a company from Virginia specialized in phytoremediation, acquired the rights to commercialize the fern by signing an exclusive license agreement in 2000 with the University of Florida which patented the use of the fern in phytoremediation. In the US, seven or eight similar companies in phytoremediation were already in existence in 2002, where the value of the potential market for phytoremediation was estimated at $100 million.

Bioremediation and Genetic Engineering

The molecular basis of heavy metal accumulation is being studied with a view to transferring the relevant genes to plant species having a wider geographic and ecological distribution. Transgenesis applied to phytoremediation is certainly incipient.

Its application on a large scale is confronted with the evaluation of risks relating to the transfer of the bacterial transgenes to plants consumed by herbivorous animals that might acquire the property of hyper-accumulating toxic metals or compounds.

Genetic transformation of the microorganisms involved in bioremediation could enhance the process through the introduction of genes controlling specific degradation pathways; it also aims at degrading recalcitrant compounds such as pesticides and other xenosubstances.

A team of US researchers led by Richard Meagher of the University of Georgia, Athens, were able to introduce into the genome of Arabidopsis thaliana, two foreign genes from Escherichia coli for the synthesis of two enzymes: one which catalyzes the transformation of arsenate into arsenite, the other which induces the formation of a complex with arsenite, that is retained in the leaves.

These remarkable results led to transgenic Arabidopsis plants which could accumulate three to four times more arsenium than normal plants and could become very effective depolluting agents. The same research team is well known for their "transgenic" work particularly in the transfer to plants of bacterial genes, coding for the conversion of toxic mercury into its less dangerous volatile form.

In Cambridge, United Kingdom, Neil Bruce and his team are transforming tobacco plants with bacterial genes that control the synthesis of an enzyme detoxifying TNT (trinitrotoluene). Other researchers are studying the degradation of trichloroethylene by transgenic plants.

ILLUSTRATIVE EXAMPLES

Thailand

Thailand provides a good illustration of the endeavours made by developing countries towards strengthening activities in environmental biotechnology and related areas. Through its Biochemical Engineering and Pilot Plant Research and Development Unit (BEC), set up in 1986 as a cooperative action between the National Centre for Genetic Engineering and Biotechnology (BIOTEC) and King Mongkut's University of Technology Thonburi (KMUTT). It has become one of a number of independent satellite research units of BIOTEC and this set-up enabled the researchers to work with less bureaucratic constraints.

During its second phase (1995-1999), BIOTEC became an autonomous organization and a large number of their researchers trained overseas and returned to Thailand to work with BIOTEC's satellite units.

Also during this period, BIOTEC assigned university staff to administer the units. BEC's first five-year plan was implemented during this second phase, with emphasis on the collaboration between researchers from BIOTEC and KMUTT; this approach to shared resources proved a successful innovation for implementing research programmes in Thailand, producing a win-win situation.

The third phase (2000-2004) was implemented in four major specific areas: algal biotechnology, sensor technology, microbial bioprocess development, and waste utilization and management. The scope of BEC covers high quality research and development (R&D), technical services, technology transfer and consultancy, in order to support and strengthen the small and medium-sized companies' competitive position.

In cooperation with the Industrial Park Centre, Pilot Plant Development and Training Institute (PDTI) and School of Bioresources and Technology at KMUTT, BEC carries out its research work while being attentive to the demand pull (i.e. technology needs from industry) and the supply push (i.e. the innovation/knowledge provided by the research institutions).

BEC's funding comes from various sources: KMUTT, BIOTEC, research grants, contract research, technical services and consultancy, training, pilot-plant design, incubator hiring and analytical services. In the 2003 fiscal year, total input from KMUTT, BIOTEC and others were approximately 25.0, 17.5, and 20.6 million baht, respectively. In the area of waste utilization and management, research projects included:

- Development of a kinetic model for anaerobic degradation;
- Development of a hybrid reactor combining an upflow sludge bed and a fixed bed in treating wastes and producing biogas;

- Population dynamics study of bacterial communities in an anaerobic hybrid reactor;
- Benchmarking of water utilization in tapioca (cassava) starch industry.

High-rate anaerobic fixed film technology (HR-AFF) for agro-industrial wastewater treatment and energy recovery has been developed up to the design and operation of an industrial-scale pilot plant. The advantage of HR-AFF compared with conventional systems lies in the fact that the bioreactor can handle large quantities of wastes per unit volume and produce biogas at a much faster rate.

As a consequence, both the capital investment and operating costs of the process are lowered, resulting in a more economically viable system. The HR-AFF bioreactor can be used with low and high-strength organic wastewater and has been proven effective for several agro-industries, such as a rice-starch, tapioca-starch factories, and vegetable and fruit-canning plants.

With respect to technical performance, the HR-AFF bioreactor can handle an organic loading rate at 2-10 kg COD/m3/day. The biogas yield was 0.4-0.5 m3/kg of COD removed which contained 60 per cent to 70 per cent methane. The efficiency of COD removal was 70 per cent-80 percent.

The HR-AFF system applied to the treatment of tapioca-starch waste-water could produce 25,000 m3 biogas/day, which could meet all the energy needs of the factory, normally met by 9,000 litres of fuel, and the rest of biogas could be used to generate electricity for sale back to the power grid of about 4,680 kWh/day.

Another achievement of BEC was the water and energy audit of tapioca-starch factories aimed at establishing a data-base and benchmarking of tapioca-starch production in Thailand for performance and efficiency improvement, and enhancement of trading potential in the global market. Five tapioca-starch factories were initially audited and the data obtained included: type of production, production capacity, and duration of operation, water and energy consumption. A second phase of the project targeted 20 factories.

The technology-transfer services provided by BEC to the industry include the following:

- Design, construction and operation of an industrial-scale anaerobic fixed-bed reactor for waste-water treatment and biogas production for a rice-starch factory belonging to Bangkok Interfood Co. Ltd.;
- Start-up anaerobic waste-water treatment plant and training for monitoring the system for a soft-drink manufacturer/bottler (Green Spot Company) since the company was able to run their waste-water treatment plant at full capacity with a high organic removal efficiency of 70 per cent-80 per cent; in addition, the cost of electricity to run the plant was considerably reduced;

- Pilot-plant fermentation services applied to wastewater treatment in the shrimp-farming industry;
- Contract research with Trangsure Co. Ltd. (frozen seafood and seafood canning factory) to improve the anaerobic sludge digester for treating excess sludge from their activated sludge systems and grease sludge from the dissolved air floatation unit.

Among the other achievements of BEC related to algal biotechnology and microbial fermentations, is the cultivation of Spirulina, which was developed from local strain selection first thru pilot studies and later, for industrial-scale cultivation.

The research involved the use of tapioca-starch wastewater as substrate for Spirulina cultivation in order to reduce production costs; the harvested algal slurry, once sun dried, contained 55 per cent protein and 7 per cent moisture. The operating costs for a plant covering a total production area (raceway-type ponds equipped with paddle wheels) of 1.5 hectares with a productivity of 40 tons of biomass per annum was estimated at between $6,000 and $7,000 per ton. The technology was transferred to the private sector where dried Spirulina is produced as high-protein feed for shrimps and fish. There seems to be good market opportunities for Spirulina biomass as food supplement for direct human consumption, in addition to its use as a feed supplement for animals; consequently the improvement of mass cultivation of the microalgae and its cost-effectiveness are important areas of research and development in Thailand and South-East Asia.

BEC has also done research on yeast. Local strains of baker's yeast and Bacillus subtilis with high yield have been selected and cultivated in a wide range of bioreactors from 5 litres up to 1,500 litres. Using locally-designed pilot-plant fermentation facilities, BEC has allowed potential private-sector clients to carry out small-scale production runs for feasibility, testing, cost predictions or optimization of processing steps, and production of prototype products for marketing tests.

These projects with private sector clients included: a pilot-scale production of pressed yeast for Bioman Co. Ltd.; a probiotics market trial for Bacillus subtilis for use in the animal feed industry for Grand Siam Co. Ltd..; and cooperative research has been undertaken with Biowealth Co. Ltd. to conduct fungal fermentation for enzyme production applied to probiotics in the animal feed industry.

Australia

In Australia, in 2004, there were more tan 60 Cooperative Research Centres (CRCs), which bring together researchers, and research users in industry, universities and government to carry out collaborative research in new, cross disciplinary areas.

The CRCs that were working in fields as diverse as agriculture, mining and information technology, received about 25 per cent of their funding from the Australian government. Previously, on 30 October 2003 at the University of New South Wales, the Environmental Biotechnology Cooperative Research Centre (EBCRC) opened with research nodes in Brisbane, Perth and Adelaide, as well as two university sites in Sydney. It was a joint venture between industry, universities and government bodies.

It will receive about A$19-million government funding spread over seven years. Cash and 'in-kind' contributions from the joint venture participants and others sources will bring the centre's total budget over its programmed life to about A$60 million.

Companies and research organizations that join EBCRC are required to make annual cash contributions and may also make 'in-kind' contributions in the form of facilities and staff. Experience has shown that the return on investment to industry members of CRC can be more than six times their investment in the CRC.

Industries that would benefit from membership include aquaculture, brewing, chemicals, food or paper processing, remediation, and oil refining and processing. The company will hold intellectual property arising from the EBCRC's research programme. Research results will be commercialized through spin-off companies, or participants in the EBCRC taking licenses for the new technologies, or direct third-party licenses.

Research projects have been selected to meet the following criteria: low to medium technical risk, time and cost constraints, and having significant potential return-on-investment. Five research projects or 'themes' have been approved to go ahead and two more have been identified for further funding namely: bioproducts from purple photosynthetic bacteria using wastes; rapid pathogen-detection strategies for the environment; novel biofilm control strategies; novel coatings for biofilm control and bioremediation; on-line process control and sludge population optimization for efficient nitrogen and phosphorus removal in high-strength wastes.

The overall goal of the EBCRC's work is to convert the by-products of industrial processes, currently regarded as waste, into high-value materials. For instance, meatworks (and the herders) obtain almost all their revenue from just half of each beast processed – the meat.

Biotechnology could convert the remaining parts into biopolymers that could become biodegradable plastics or valuable chemicals. Breweries and food processors also produce wastes containing materials suitable for biotransformation. Outcomes of research projects will include:

- New fast, low-energy bioprocesses producing high-value products.
- Chemical precursors for manufacturing, e.g. esters, polyhydric alcohols (glycols).

- New biosensors and control systems.
- Purification and separation processes for chemicals.
- Reduced greenhouse gas emissions.
- The EBCRC also offers scholarships to students to undertake higher degrees and to develop the necessary research expertise across the full range of the centre's research. Students are provided with opportunities for work experience during their studies. Post-doctoral fellowships are also offered to attract skilled researchers, especially expatriates.

Through its research and training programme, the EBCRC aims at deeply changing the way in which Australia deals with waste and at showing how biotechnology can contribute to sustainability – whether in biofilms, bioproducts, biodecontamination, biofuels, or biotransformation, or just enhanced bioprocesses using the natural ecosystems.

METAGENOMICS

Craig Venter, the US biologist known for leading the private endeavour to sequence the human genome, is attempting to catalogue the microflora of oceanic environments. Onboard the Sorcerer II, it has already sailed along the east coast of North America, through the Panama Canal and around the Galapagos Islands. Every 200 miles, onboard researchers haul up 200 litres of seawater and filter it for microbes.

Samples are frozen and sent to the USA, where biologists at the Institute for Biological Energy Alternatives in Rockville, Maryland, will try to sequence the DNA of every organism in each sample.

The expedition has the highest profile among many other similar projects under way around the world, in environments ranging from geographical (abandoned mines) to physiological (the human gut.)

After sequencing the DNA of single species (there are complete sequences for more than 300 organisms), researchers are trying to sequence the DNA of living species in defined environments – a science called environmental genomics, or metagenomics.

Through better understanding of how myriads of bacteria combine and interact to influence the ocean, we shall be better able to monitor the conditions of these environments and possibly manipulate them to our advantage. It may even be possible to find micro-organisms that can produce drugs or act as energy sources.

Science is now able to grow 3,000 species of bacteria on artificial culture media, but the vast majority – probably more than 99 per cent – resists such attempts. It seems impossible to approach such a huge microbial diversity with conventional techniques. Metagenomics can access this information.

The principle is simple: select an environment; scoop or scrape up a sample for it; and sequence the genes in that sample. Lab researchers can look for diagnostic DNA sequences to count the number of species in a sample; they can put DNA fragments into bacteria that can be grown in the laboratory, and screen them for properties such as antibiotic production; or they can simply pull out sequences at random to give a picture of the number and type of genes in that environment. In practice, however, there is too much diversity in most environments to sequence every gene of every species.

Craig Venter's team collected a sample of 1,500 litres of water from the Sargasso Sea, in the Atlantic for their global study. They chose the Sargasso Sea because it was presumed to be poor in nutrients, and was thought to be one of the ocean's less diverse regions.

But to their surprise, they found 1,800 species, including 150 new to science, and 1.2 million new genes.

Environments richer in nutrients, such as the soil, will probably be even more diverse still. The pilot study in the Sargasso Sea cost $2 million; C. Venter estimates that a more complete scan would currently cost five times that. The US Department of Energy funded the Sargasso Sea project. Thorough surveys of such places will have to wait for expected falls in the cost of sequencing.

Metagenomics could help us understand environmental problems such as carbon dioxide take up by bacteria to reduce its concentration, the breakdown of pollutants by micro-organisms, It could also help to detect environmental damage, because once we know what it is that is out there, it is a step to be able to determine any changes to the environment.

Biotechnology companies such as Diversa, based in San Diego, are also working to mine the data for useful genes. A new antibiotic, turbomycin, was recently discovered during a metagenomic scan of soil by a team led by Jo Handelsman at the University of Wisconsin, Madison. For the time being, C. Venter's main objective is to gather information: 'the main purpose is cataloguing what is here and the diversity is higher than most people imagined; we are spending billions looking for life in Mars, but we do not have the slightest idea of what life there is on this planet'.

SOCIAL ACCEPTANCE

The potential of 'white' biotechnology is conducive to the tie-ups and linkages among government, industry and academe to work within a sustainable development framework.

By providing alternatives such as new materials and fuels that are not derived from petrochemical processes, by improving and enhancing the bioremediation of water, soils and ecosystems at large, by trying to use less fossil-fuel energy, 'white' biotechnology will enjoy a positive social acceptance, even among environmentalists, if it continues on this path.

However, a crucial factor in social acceptance particularly by environmentalists is the conclusive impact on biodiversity of releasing into the environment genetically-modified organisms used in the processes of 'white' biotechnology, including bioremediation; and this is still largely unknown. Industrialists use "white biotechnology" mostly in confined environments, such as their factories, bioreactors and greenhouses, while applying strict biosafety regulations. Until then, widespread social acceptance may have to wait.

APPLICATIONS OF ENVIRONMENTAL BIOTECHNOLOGY IN INDUSTRY

Biotechnology is the application of biological systems in technology that can only be achieved through an integration of the biological, physical and engineering sciences. The current approach of biotechnology, in the forest products industry is to apply biological products such as enzymes rather than live cultures.

Enzymatic action is very specific and also easy to control. This paper considers biotechnological applications in different operations including forestry, wood preservation, pulping, bleaching, deinking, papermaking and water treatment. The benefits of biotechnological applications are: improved product quality, production rate or diminished environmental impact. Utilisation of enzymes will become more widespread as biological products become less expensive through large-scale production.

Biotechnology has been defined as the integrated use of biological, physical and engineering sciences in order to achieve technological application of biological systems. The goal of technological application implies that biotechnology excludes fundamental research and that it must be relevant to industry. Fundamental studies do, however, play a crucial role in the development of the applicable biotechnological processes. Recombinant DNA technology can, for example, be used to exploit or enhance the properties of natural biological systems for industrial purposes.

Most biotechnological processes make use of microorganisms such as bacteria, yeasts and filamentous fungi, but vascular plants, algae and even animal tissue can also be utilised. Classical biotechnological processes have been used for ages in the production of food and beverages such as bread, cheese, beer and wine.

These processes are characterised by the direct application of live organisms and the in situ production of enzymes and other products. New biotechnology has a stronger focus on the application of biological products or enzymes and these are often produced ex situ. The advantages of the new biotechnology over classical biotechnology are that mass balances are easier to solve and process control is improved. The forest products industry is based on the processing of biological raw material and it is, therefore, well suited to the introduction of biotechnology. Biotechnological processes have been developed for most of the unit operations of the industry and a number of them are commercialised. The aim of this paper is to examine the potential of different biotechnological processes and their integration into conventional operations of the forest products industry.

Biotechnology Applications

Forestry

Biotechnology can be applied in forests and especially plantations to improve fibre production and to protect trees. When exotic tree species were initially planted in South Africa, it was observed that stunted growth occurred due to the absence of mycorrhizal symbionts. Soil, containing these fungi, was then introduced into the country to improve growth of trees. Today, it is believed that fibre production can be improved by inoculating seedlings with more efficient mycorrhizal fungi. Mycorrhizal fungi enhance plant growth through improvement of mineral and water absorption, protection against pathogens and even secretion of growth enhancing hormones.

Tree protection can be achieved by using biological control agents such as fungi or bacteria. The biological control agents (BCAs) are most effective against root diseases and especially where they are applied in an enclosed environment such as seedling trays. These BCAs act by competing with pathogens for nutrients and space, by secreting toxic substances or by direct consumption of the pathogens.

Treatment of Timber

After felling, logs can be treated with fungi to improve debarking or to protect the wood against blue staining and decay. Inoculation of softwood logs with the white-rot fungus Phlebiopsis gigantea leads to rapid colonisation of the logs during transport or storage. The fungus is able to loosen bark, reduce pitch and protect the wood against sap staining. The loosening of bark is accomplished during the degradation of cambial cells by the fungus. The best-known biological agent to protect timber against blue staining is the fungus Ophiostoma piliferum. A white mutant of this fungus has been selected and the inoculum is produced commercially as Cartapip. The fungus is a primary coloniser of wood and is able to utilise sugars and extractives in the wood without affecting the strength of the wood. These characteristics enable the fungus to out-compete any sap-stain fungi that subsequently infect the wood.

Biopulping

Biopulping is a solid-substrate fermentation (SSF) process where lignocellulosic materials are treated with fungi prior to pulping in order to reduce energy during mechanical pulping or to reduce chemical consumption during chemical pulping processes. Research has focussed on the utilisation of lignin-degrading fungi for biopulping, but the only commercial process currently available, utilises O. piliferum. Wood chips have also been treated successfully with fungal enzymes to improve the penetration of pulping liquor.

The aim of bio-thermomechanical pulping is to reduce the energy consumption during pulping and to improve pulp strength. The beneficial effects of fungal treatment on mechanical pulping can be ascribed to the reduction of the binding capacity of fibres and improved fibrillation. During bio-chemical pulping, the lignin content of the pulp or pulping time can be reduced. These benefits are the result of increased lignin solubility that is caused by fungal degradation.

Depitching is one application of biopulping where Cartapip is applied to stored wood chips and the pitch in the wood is reduced by the fungus metabolising it. It is evident that different fungi are suited to different pulping processes, but these processes must be adapted to achieve the full potential of an environmentally friendly technology.

Pulpwood Preservation

Cost of wood contributes more than 80 per cent to the variable cost of unbleached chemical pulp. Contaminating organisms that degrade cellulose or cause staining of wood can, therefore, have a significant impact by reducing yields of chemical pulp or necessitating bleaching of mechanical pulp. Biopulping fungi are inoculated at high dosages and incubated under conditions that favour their ability to colonise and compete with undesirable organisms. These fungi are thus able to protect pulpwood against decay and staining organisms.

Biobleaching

Two different approaches have been used to enhance bleaching through enzymes. The first method utilise hemicellulolitic enzymes such as xylanase. These enzymes are unable to degrade lignin, but cause limited degradation of hemicellulose in kraft pulp to expose lignin to further attack by bleaching chemicals. Xylanase pre-bleaching is used at several kraft mills worldwide. In 1994, eight per cent of Canada's bleached kraft pulp production was treated with xylanase. This technology holds environmental and economic advantages, especially since the enzymes have become less expensive as result of increased market size.

The typical chemical saving associated with the use of xylanase is more than 15 per cent active chlorine. Unfortunately, the limit of approximately 20 per cent in chemical savings cannot be exceeded. The second approach to enzymatic bleaching is the utilisation of oxidative enzymes, mainly laccase and manganese peroxidase (MnP) to delignify pulp. Of these enzymes, laccase is preferred, because MnP requires hydrogen peroxide, manganese (II) ions and a chelator. Laccase can cause delignification of pulp under slight oxygen pressure, but is considerably more effective when mediators are added. The most important obstacles to commercial application of laccases are the lack of sufficient enzyme stocks and the cost of mediators. Marked progress has been made over the last two years to solve these problems and it is expected that biobleaching with laccase will be able to compete with other TCF bleaching processes.

Enzymatic Deinking

Enzymatic deinking can be applied on old newsprint and office waste, but it is more effective on the latter, because it contains large amounts of laser and xerographic inks that are more difficult to remove by conventional methods. The commercial cellulases used, are able to hydrolyse fines to release ink particles. The ink can then be removed by chemical or air flotation. A secondary benefit of the enzyme treatment is that the stripping of fines from fibres results in improved drainage without reduction in strength properties.

Improvements in Paper Making

Most of the applications of enzymes to improve the papermaking characteristics of pulp are aimed at the improvement of secondary fibre. Application of cellulases and hemicellulases to modify fibres has resulted in the following benefits·

- Improved beating to save energy;
- Higher freeness;
- Enhanced drainage;
- Improvement of certain strength properties;
- Decreased disintegration time of recycled pulp, and
- Reversal of hornification.

In other applications, enzymes are used to remove the non-fibrous fractions of secondary fibre that hamper drainage. A good example of such material is residual starch that is especially abundant where old corrugated containers are used as raw material. Amylases can be applied to degrade remaining starch that would not contribute to the web characteristics, but would restrict drainage. In one mill, application of amylases resulted in a 6,9 per cent increase in production of paper.

Control of Microbial Fouling

Microorganisms in mill water systems cause slime or biofilm build-up, odour and corrosion. The consequences of the contamination are increased web breakages, holes and spots in paper as well as increased cost of maintenance.

Biocides are generally used to control microbial build-up, and deposits are removed during boilouts by sodium hydroxide or dispersants. Inoculation of water systems with competing microorganisms or bacteriophages have been proposed, but the most feasible biotechnological approach appears to utilise enzymes to prevent aggregation of biofilm-producing organisms and degrade extra-cellular polysaccharides. These enzymes include proteases and carbohydrate-degrading enzymes and combinations of these enzymes are often applied. The enzymes also expose microorganisms to toxins by reducing protective polysaccharides, and are usually dosed in combination with biocides.

Effluent treatment

The oldest application of biotechnology in the pulp and paper industry is probably that of wastewater treatment. Biological processes are usually employed in secondary or polishing treatments that follow sedimentation or other primary treatment. Biological treatment processes include: Aerated stabilisation basins, activated sludge, oxygen-activated sludge, trickling filters, rotating biological contactors, anaerobic lagoons, upflow anaerobic sludge blankets, anaerobic filters and anaerobic fluidised beds. These systems are all characterised by a complex microbial community that is responsible for the improvement of water quality.

Biotechnology is applied on a commercial scale in many operations of the forest products industry. Application of enzymes has become very prominent recently, because they are highly selective in their action and have a negligible environmental impact. Wider application of these enzymes are restricted by cost and availability of the enzymes. This problem will soon cease to exist, given that the cost of enzymes has decreased considerably over the last decade as the market increased. A number of companies are now also producing enzymes specifically aimed at the pulp and paper market. One remaining hurdle, is the perceived incompatibility of biotechnology with conventional processes. This problem is more difficult to solve since it is largely due to a lack of understanding of biological systems and a lack of acceptance that biological products are expected to work under extreme conditions in mills. It is, therefore, clear that biotechnology can only be developed through integration and understanding of biological as well as physical sciences.

INDUSTRIAL ENVIRONMENTS OF POSTWAR AGRICULTURE

Once policymakers began to administer a system of minimum and maximum prices for corn, hogs, and other commodities, however, it was very difficult for the price of corn and the price of hogs ("the corn-hog price ratio") to stay in balance. During World War II, secretary of Agriculture Claude Wickard aggressively encouraged hog production through price guarantees, and American hog production hit new records in each year from 1941 to 1945.

As a consequence, however, supplies of feed grains fell sharply. This compromised plans to use corn and other grains for production of industrial alcohol, synthetic rubber, and other wartime needs. Shortages of cottonseed meal, fish meal, and other protein supplements further challenged hog producers. In 1943, in the face of record worldwide demand for the American food products, the USDA launched a voluntary Protein Conservation Programme that encouraged the use of vegetable proteins as an alternative to animal proteins.

This crisis turned out to be the issue that launched Damon Catron's research career in the realm of swine nutrition and evidently sparked his interest in maximizing feed efficiency. In all, the "feed famine" and other lessons of World War II taught hog farmers that they were increasingly dependent upon commercial feed supplies and international events beyond their control, and that pressures of industrialisation were sure to impact the typical hog farmer. As other historians have noted, the war removed the "conceptual blocks" that had limited the application of scientific and industrial innovations to agriculture, which encouraged the adaptation of "scientific" values as appropriate for nearly all social problems.

Circumstances in the agricultural labour market also pushed the redesign of the American hog enterprise along industrial lines. World War II brought intense farm labour shortages; the situation did not improve significantly following the demobilisation of the armed forces. In 1943 researchers at Purdue University founded the Work Simplification Laboratory, an endeavor that aimed to reduce human physical labour on the farm. Specialists in hog management used stopwatches, video cameras, and balls of string tied to the farmer to evaluate the time and effort "wasted" in climbing over fences, opening and closing doors, and hauling feed and water. In reports that purported to apply the science of human dynamics and the practices of industrial management to farm operations, researchers claimed farmers could easily save one-fourth to one-half of their labour.

Significantly, the work simplification specialists strongly promoted the relatively innovative notion of multiple farrowing-their tables demonstrated that a second litter annually was worth the investment in time and energy. Stagnant demand for pork products and byproducts created yet another incentive for hog producers to streamline their industry and view their resources as commodities. Traditionally, American meat packers and butchers utilised hogs as much for their byproducts as for their meat.

Most pork meat products were not sold fresh but processed through salting or curing. Farmers took hogs to market on rather indeterminate schedules and at a wide range of weights. At times when the corn-hog price ratio was favourable, as in World War II, price supports encouraged farmers to fatten their animals indefinitely, and they often sent a three-hundred-pound porker to market. This approach faced challenges in the postwar period, when new vegetable oils and detergents weakened consumer demand for animal lard, and postwar consumers found the consistent supplies, prices, and quality of chicken, turkey, and frozen fish preferable to pork products. Invented not by "the producer or processor but rather by the housewife, " the "meat-type" hog emerged as the new standard, a leaner animal that could be marketed routinely at about 220 pounds. With commercial feeds replacing homegrown grains and scraps in hog rations, it became easier for farmers to raise their hogs consistently and reliably to the ideal of this standardized and industrial organism. Domestic and global politics provided yet another incentive to industrialize hog production methods. Erratic swings in meat supplies and meat prices proved a continual thorn for the Truman administration, particularly as a "deep...meat famine" and threats of a producers' strike appeared on the eve of the 1946 elections. American policymakers also embraced the use of farm products as a component in their suddenly internationalist agenda in foreign affairs.

Improving agricultural production became part of what one historian has labeled the "technocratic front in the Cold War" as political and industrial leaders urged farmers to take stock of their role in the global crises. These geopolitical considerations reached the local level when farm conventions and journals began delivering the messages that Soviet spending on industrialisation could outpace Americans', and that farmers needed to be vigilant of the possibility of enemies striking the nation's food supplies through bioterrorism. Journalists also reported that the demographic pressures of the baby boom generation, combined with America's suddenly expanding role in world affairs, were placing new demands on farm production.

A 1946 article in *American Magazine,* for example, described the techniques of "genetic engineering," "redesigning nature," and "custom-built livestock" being developed at the USDA research facilities in Beltsville, Maryland. Its author, Frederick Brownell, also recognised the political implications of these new foods. It was time for Americans to repay other nations that helped win the war, he argued, and to fulfill Franklin Roosevelt's promise to provide "Freedom from Want." Similar rhetoric appeared in Maxwell Reid Grant's 1949 essay "Engineering Better Meat" published in *Popular Mechanics.* Employing the industrialists' and engineers' language of "blueprints," "specifications," and "model years," Grant explained that "nature needs help" if Americans were going to rebuild Germany and "feed a hungry world." The sum of these numerous background issues led to a fairly self-evident conclusion for the farmers and industrialists involved in swine production: they had a mandate to reshape both the hog industry and the hog animal in ways appropriate for the postwar era.

It was in this context that Damon Catron of Iowa State began to articulate his interest in industrialising agriculture. In an address delivered in the fall of

1946, Catron explained that Iowa's hog farmers needed to buy "a system of feeding" rather than just a bag of feed. He linked that to changes in the hog organism, for one could not have "poor machinery to put the raw product [or feed] through." Yet his speeches that autumn did not reveal an easy way to reach that goal. In this era, animal husbandry experts typically focused their research on three branches of their science: sanitation, breeding, and nutrition. But influenced by his University of Illinois colleagues who called for a "new fundamental philosophy" in the interpretation of swine feeding, Catron's lectures urged further research on ideal ratios of nutrients, such as fats, proteins, carbohydrates, minerals, and vitamins, a departure from the previous emphasis on searching for an ideal ratio of ingredients, such as corn, alfalfa, milk, and the like.

Yet Catron also admitted that his scientific colleagues did not have answers for many of the fundamental problems that Corn Belt hog farmers faced. The solution, Catron suggested, fell under the final theme of his speech, "The Need for College-Industry Cooperation." Although swine industry observers readily recognised the many incentives to alter hog production practices, few imagined that pharmaceuticals would be at the centre of dramatic change. Then antibiotics burst suddenly upon the scene. Led by Thomas Hughes Jukes and E.L.R. Stokstad of the American Cyanamid Company's Lederle Laboratories, researchers discovered in the late 1940s what came to be known as the "antibiotic growth effect"- that the feeding of antibiotics at low levels to agricultural animals resulted in enhanced growth.

This research had its origin in a twenty-year search for the animal protein factor (APF), something that seemed to be a combination of vitamin-like substances that could raise the efficacy of feeding animals cheaper plant proteins to the level of commonly used animal proteins. Indeed, Catron and others already had worked with the Lederle laboratory in establishing that farm animals did not gain weight at satisfactory rates when fed vegetable proteins alone.

There was something important in animal byproducts, yet whatever it was remained undetected. For a time, vitamin B12 seemed to be the mysterious APF factor, and manufacturers at Lederle, Pfizer, and other major pharmaceutical firms produced feeds fortified with Vitamin B12 for the 1949 season. Searching for a method that would permit production of the new vitamin on an industrial scale, manufacturers found a ready supply in residues from the fermentation processes used in the production of the antibiotic wonder drugs aureomycin and streptomycin.

Meanwhile, researchers began to notice connections between these common antibiotics and animal growth. Stokstad and Jukes found that vitamin B12 produced from the antibiotic organism yielded more rapid growth in chicks than a pure vitamin extracted from liver. They first presumed that the vitamin was the principal cause of the weight gain, but then deduced that the antibiotic alone might be beneficial to the chicks' growth. A second test with antibiotic feeds on piglets produced even more spectacular results. Stokstad and Jukes quickly published two studies in early 1949 announcing the antibiotic growth effect. That year, American Cyanamid sent samples of antibiotic fermentation materials to prominent animal scientists at several land grant colleges and experiment stations. According to surviving documents, Lederle's scientists did not allude to the emerging significance of the antibiotic growth effect. For example, the company's $1,500 contract with Tony Cunha, a researcher at the University of Florida, suggested that his research would be on folic acid and vitamin supplements, even though the publications that emerged from this study were perhaps the first independent studies confirming the antibiotic growth effect.

Lederle's agreements with Catron at Iowa State were similarly vague. After taking Catron and his wife to see the Rockettes in New York, Lederle representatives offered Catron a $1,500 contract to study the animal protein factor in various hog feeds. By December 1949 Catron could see that the pigs thrived on rations fortified with the antibiotic aureomycin, but he did not know the formula and thus could not be sure why growth had accelerated.

"I would certainly like to know more about this product with which I am working," Catron told his corporate sponsors. In any case, the studies conducted by Catron and Cunha in 1949 provided the independent confirmation necessary for Lederle to announce its breakthrough discovery.

The public impact of this discovery became evident in April 1950, when Stokstad and Jukes presented their results at the American Chemical Association meeting in Philadelphia. The *New York Times* placed its story on page one under the headline "'Wonder Drug' Aureomycin Found to Spur Growth 50 per cent." The *Times* reported that the new drug was better than any vitamin; it could speed growth of hogs by as much as 50 per cent, with similar results for young chickens and turkeys. Early studies indicated an ever more encouraging picture of the antibiotic growth effect.

Even small amounts of antibiotics in feeds improved both meat and egg production in poultry. Antibiotics stimulated growth of hogs on both dry lots and in pasture; and they accelerated growth whether administered intravenously or through feed and water. Catron also claimed that medicated feeds reduced the loss of runt or weak animals and produced piglets "four times more uniform in size" than those raised under natural conditions.

Early studies also indicated that antibiotic feeding reduced disease among animals, whether administered at therapeutic or subtherapeutic levels, and with no undesirable side effects. Apparently, an agricultural panacea had entered the public arena.

Unsurprisingly, this issue quickly swept the agricultural press in 1950 and 1951. Dozens of researchers published evidence corroborating Stokstad and

Jukes's claims. Experiment station and university scientists produced reports that demonstrated the action of antibiotics on the growth of chicks, turkey poults, and hogs; less conclusive experiments on lambs, dairy calves, and steers were also published in that year.

Signaling the greater importance of disseminating their findings directly to producers rather than reaching researchers through academic journals, Catron, Cunha, and Jukes regularly spread the gospel of antibiotic feeding before industry groups like the Kansas Formula Feed Conference, the International Baby Chick Association Meeting, and scores of similar state and regional meetings attended by feed dealers, extension agents, agricultural journalists, and other representatives of the emerging agribusiness complex. The Food and Drug Administration (FDA) was called in to investigate the claims. Within a few months, and with only moderate scrutiny, the FDA ruled that antibiotics could be advertised on animal feeds beginning in January 1951. Damon Catron and Iowa State College seized upon these findings and other postwar opportunities to build the academic-industrial links that he had envisioned. In 1949, for example, Catron could boast that the state legislature had been "good to us" since it had appropriated funds for new swine research facilities and test pastures.

In 1951, Catron secured grants of $3,000 from Pfizer, $2,500 from Merck, $5,000 from American Cyanamid, and $20,000 from a number of other pharmaceutical, chemical, and agribusiness firms to fund the new Centre for Swine Nutrition Research. Catron welcomed these changes as appropriate and necessary. Through more systematic hog-raising schemes, he also believed the enterprise could become more specialised, and hoped that professional swine producers would take over the industry, thereby removing the "in and outers" who raised hogs only on a sporadic basis. Catron also vigourously promoted antibiotic feeding in the context of cold war politics. In a report intended to speed federal regulators' approval of antibiotic production methods, Catron recalled the "critical shortages" of feed that hampered the war effort in 1944. In response to the "national emergency" of 1951, Catron cited the power of antibiotic feeds to reduce human labour and to utilise relatively inexpensive vegetable proteins as justification to accelerate production.

Although crowding animals together naturally increased their susceptibility to disease, scientists understood that antibiotics and other pharmaceuticals could overcome that natural burden, or at least make it possible for fewer farmers to handle more animals as part of the process. The major pharmaceutical and feed firms seized upon the success of antibiotic feeding to promote the American farmer as a participant in a modern industrial system. The Lederle firm, for instance, issued a booklet that used simple cartoons to contrast the ragged clothes, limp cigarette, and 1920s-era jalopy of the farmer who eschewed commercial feeds with the starched clothes, fat cigar, and full-sized automobile of the farmer who embraced the products. Throughout the booklet, the text emphasised that commodities reduced the labour required of the farmer, for much of the work had been transferred to the animal. Meanwhile, Catron collaborated with Pfizer and the American Farm Institute to produce the "Open Door to Greater Hog Profits," a film that offered an informative illustration of the emerging links between antibiotics and interventionist production methods. The film began with Catron's ten-minute lecture on feed rations. Among other lessons, Catron reported that runt pigs fed on antibiotic feeds yield a 54 per cent improvement in growth rates. Then, the film portrayed an old, haggard, disinterested female farmer, bedecked in cheap clothes and ragged scarf, shoveling corncobs and other crude rations into a hog pen. The producers contrasted her inefficiencies with the capabilities of a cleanly dressed male farmer, who was depicted reading a farm journal in front of a real symbol of modernity-a television.

Ideal farmers were aware of the latest research and determined to apply the rather self-evident lessons of using scientifically balanced rations, tending to hygiene and sanitation, and trying to save every piglet upon farrowing. In essence, this film captures the moment of transition from traditional to modernist methods of hog production, highlighting the potential of the still experimental confinement hog operations.

The film also repeatedly showed a sight that has been quite rare in rural America for decades-isolated hogs feeding themselves in lush pastures as if they were on the open range. Catron next sought to link improved feeding efficiencies with the goal of year-round farrowing in order to create a more steady supply of market hogs. The discovery that baby pigs could be weaned from the sow within just a few days of birth if given feeds fortified with antibiotics and vitamins to replace their mother's milk further expanded the producers' control over their pigs.

The resultant physiological changes meant that the sow could be bred again within just nine weeks of giving birth, thus drastically altering both the natural reproductive patterns and breaking the traditional animal cycle linked to the corn harvest. As meat packer Jay C. Hormel explained, the "sow might immediately be put back to work producing another litter instead of performing no other service than milking her litter."

"Multiple farrowing" combined with more efficient feeds meant that hogs could reach market weight within five or six months of birth, thus turning Midwestern farmers into producers of a virtually continuous flow of finished products. To further regulate and standardize animal feeding practices, Catron's next area of research was the notion of "life cycle feeding." According to this principle, pregnant sows, lactating sows, piglets, growing pigs, and fattening hogs each required different and sophisticated feed formulas. In March 1954 Catron introduced the "I.S.C. Pig Pre-Starter '75' Formula, " a complex feed mixture intended for the crucial stage in young pigs' lives after their unnaturally early weaning from the sow. For Catron, the formula provided a crucial tool that enabled him "to study the nutritional requirements of baby pigs without interference from the sow." The formula included seventeen basic ingredients, one of which was a mixture of vitamins and antibiotics that contained another sixteen ingredients. Nevertheless, pre-starter feeds struck a chord with Midwestern farmers; in one four-day period, Catron received 1,500 requests for information about the formula and how it could be obtained. Catron made arrangements with several feed companies to sell his formula to the public; he even urged his brother, a farmer in rural Indiana, to jump on the opportunity to invest in this branch of the business.

By insisting that local farmers could not "do a good job of mixing their own pig starters" Catron again contributed to the commodification of the industry. Yet it is also significant that the pre-starter formula defined the point of resistance between the microlevel research of the scientists and the macrolevel production goals of commercial feed producers. Since pre-starter feeds were especially complex in their formulation but were intended only for a brief period of the animal's lifespan, feed companies balked at pressures to produce such feeds.

Meanwhile, Catron introduced a simple tool that contributed to the goal of manipulating the life cycle of the mature hog approaching market weight. The "Pork Costulator," a small, circular slide rule, enabled farmers to abandon imprecise feeding methods and calculate ideal ratios of corn and other feeds. The device had calibrations on two scales, one for "least cost" for farmers who wanted to bring their hogs to market weight at minimal cost, and another for "least time" for those who wanted to get their hogs to market quickly in order to hit anticipated favourable prices. In all, the response to pre-starter feeds and the Pork Costulator indicated that many farmers were willing to accept Catron's boast that "swine nutritionists…know more about what the pig needs than the pig himself."

Catron and other land grant university researchers continued to fit additional pieces into the puzzle of industrialising the hog organism and the hog production system. For Catron, an extensive tour in 1954 of the leading animal science research centres in Europe left him with the impression that Europeans were far behind Americans in terms of antibiotic feeding, but that they had much to teach about improving livestock housing and management practices. By the mid-1950s, compressing time and space became the central tenets of this experimental agenda, and "environment," "confinement," and "integration" replaced antibiotics and feeding as the buzzwords of swine research. In a remark that may allude to the Pfizer advertisement used to introduce this essay, Catron explained "in my opinion the farmer has waited too long for the feed man to put the miracle drug in the feed bag."

As Louisiana hog farmer Homer Harris put it, it was time to "modernize our entire hog farrowing assembly line" by lessening the time between farrowings and restricting animal mobility. In response, researchers examined countless designs for centralized hog houses, many outfitted with artificial heaters, air conditioners, automatic waterers, electric lights, and even offices. Signaling the opportunities that these trends offered to firms beyond the traditional agribusiness sector, the Reynolds Aluminum Company funded Iowa State's agricultural engineers to investigate connections between swine growth, feed efficiency, the temperature of the animal's environment, and materials used in construction of hog pen buildings.

A shift in Damon Catron's grantsmanship activity during this era offers another illustration of the trend towards the industrialisation of the hogs' environment. Explaining "we are reaching the limit in substituting nutrition for good management, " Catron presented a new seventeen-point programme of efficient hog management that called for manipulations of the swine diet, environment, breeding, veterinary care, marketing, and more.

In an appeal for funds from farm building and equipment manufacturers, Catron outlined the need for research on automated feedlot equipment and facilities designed "for comfort of the pig-and of the operator." These studies comprised Catron's programme of "life cycle housing," the notion that farmers should coordinate their "life cycle feeding" with investments in separate housing units for the distinct stages of the hog's life. Life cycle housing called for farmers to invest in air conditioning, heated floors, regulated water temperatures, germicidal lamps, and other aspects of a manipulated environment.

Manufacturers and hog producers responded first with designs for various systems intended to bring sows in from the pasture during farrowing time. Indoor farrowing and early weaning demanded new housing complexes that separated sows and piglets, and structures built with guardrails, sloped floors, and other precautions to "remove the sow hazard, " or the possibility of death by crushing. Also, early experiments with hormone treatments pointed to the possibility that one could precisely schedule the birth process to ensure that sows would give birth at times convenient to the farmer.

The emergence of confined housing systems that eliminated the need for pasture altogether proved an especially significant development. According to one analysis, multiple farrowing was the main incentive for farmers to move towards the "assembly line production" on concrete feedlots, since continual crops of young pigs would simply require too much labour to continually move fences, feed troughs, and waterers through the traditional system of using pastures. The advantages of artificial environments challenged the pasture system in other ways as well.

Researchers at Purdue, for example, reported "startling" losses of hogs on hot summer pastures, and issued reports promising that "manmade weather can improve livestock efficiency." In 1959, *Successful Farming* magazine highlighted the operations of J. Herbert Doak, an Ohio farmer who also had been featured in 1952 for his innovative use of pasture and continuous farrowing. By the late 1950s, however, Doak had shifted his operation entirely to concrete, convinced that he could manage disease threats and lessen his labour requirements in more confined spaces. Antibiotics also played a role, challenging the notion that pastures were naturally more healthful environments than confined hog lots. Although crowding animals together naturally increased their susceptibility to disease, antibiotics, other pharmaceuticals, and regularly dousing pens with lye and creosote were cheap insurance policies that helped reduce disease risks. By the late-1950s, advertisements of the Portland Cement Association, with their slogan "The Mark of a Modern Farm...Concrete," appeared alongside the advertisements for artificial housing, manufactured feeds, feeder pig companies, and "high-dosage" antibiotics.

Although converting to a system of confinement and concrete required a capital investment too great for tenants and many small farmers, many analysts argued that for others, the benefits out-weighed the costs, especially the opportunity to convert pasture into productive cropland. As powerful as the juggernaut of pressures to adopt the factory farming paradigm appeared to be, it is significant that many farmers questioned the trends. Some blamed "integration" as the root of their problems, for it seemed that packers, grocery stores, feed merchants, and pharmaceutical firms were prospering despite low profits on the farm.

Understandably, farmers objected most passionately to the impact of industrialized agriculture on commodity prices. In 1955 hog prices fell from nineteen cents per pound in May to a thirteen-year low of eleven cents per pound in December. Rural citizens from many trades and occupations wrote to President Dwight Eisenhower, Secretary of Agriculture Ezra Taft Benson, and others in government with grievances about prices, pleading to slow the pace of agricultural change. Dozens of Midwestern citizens protested the cost/price squeeze that many attributed to the expenses of pharmaceuticals, commercial feed, and new buildings and equipment.

California hog farmer Warren Jaycox sent Secretary Benson a sarcastic Christmas card depicting hog farmers begging with a tin cup. Thirteen-year old Charles Borstad of Minnesota was among the many who appealed for help, explaining that he faced losses of at least $10 per head on the hogs he had raised through his 4-H Club. Agribusiness concerns also faced pressure from the National Farmers Organisation, an aggressive grassroots organisation founded in 1955 to fight for higher commodity prices in the corn-hog belt.

There is also ample evidence that farmers did not readily adopt the Fordist, assembly line model that antibiotic feeding promised and Catron promoted. Farm journals routinely featured farmers who creatively improved their operations with inexpensive and homemade equipment. Examples include feed bins and water troughs placed on sleds that could be hauled by tractor from one pasture to another; cheap automatic feeders built from old tires and scrap metal; and walk-through corncribs that enabled hogs to feed and find shade at the same time. Other agricultural experts also encountered signs of resistance. Farm extension agents in Florida, for example, complained in 1954 that local farmers seemed ignorant about the latest recommendations in swine nutrition. Their 1956 report promised to teach more aggressively the doctrine of centralized swine management, but it noted that fatty carcasses, market gluts, and batch production schedules remained common in Florida.

In a speech at the 1956 Iowa Swine Producers Day, poultry producer Leon Johnson complained mightily of the "whining" hog producers who refused to see the wisdom of large-scale capitalism. Although limited in their impact, such cases corroborate what historian Ronald Kline describes as the "contested processes" and "individual modernities" with which rural Americans often encountered social and technical change. Yet such cases could slow development of the industrialized pig only temporarily.

By the end of the 1950s, future trends in hog production were already visible, even if their final implementation remained incomplete. Agribusiness concerns, farm journal editors, and academics from the land grant universities continued to push for the further manipulation of nature into artificial agroindustrial systems.

Antibiotics and artificial environments continued to bring consolidation and rising productivity to American agriculture in the decades after 1960. Confined operations centreed on concrete feedlots defined the new-era Midwestern farms, as hogs on pasture became a distant memory. In all, changes in the hog industry exposed growing interconnections between agriculture, technology, ecology, and political economy.

Most of the changes that occurred between 1945 and 1960 occurred at the level of what Edmund Russell has labeled "macrobiotechnologies," or the interventions that impacted the lifestyle and physiology of the whole organism. But in a speech delivered to the Midwest Feed Manufacturers Association meeting in 1960, Catron hinted that many future changes would be approaching the realm of Russell's "microbiotechnological" interventions, or those that occur at the level of cell and molecules.

Catron predicted that hog producers of the future would develop ever more precise and complex formulas for their feeds, aided in part by the "electronic brains" that some feed companies were beginning to utilise. He expected researchers to develop more specialised feed additives that would affect the animal in precisely targeted ways. For instance, predigested foods and certain enzymes might alter the hog's natural digestive function and extend feed efficiency. Other feed additives could alter and improve natural expectations for controlling estrus, ensuring embryonic survival, developing leanness of the animal carcass, and "other biological functions in the animal over which the farmer would like control."

Indeed, Catron forecast that farmers of the future could generate performance "exceeding" the animal's actual "genetic potential." Catron's speech also accurately predicted the growing attention to the genetic "plasm" of the boar and sow, to artificial insemination, and to a future of pigs being delivered by caesarian section and reared in isolated, disease-free environments. In all, in this speech and elsewhere, Catron called for farmers, like managers of a factory, to pay greater attention to strategies that would induce the animals, like laborers, to do even more work.

7With "work" in this context defined as having hogs consume the feeds that maximized growth efficiency, Catron suggested that hog producers learn to pay greater attention to environmental conditions that caused animal stress, such as uncomfortable temperatures, bothersome noise from airplanes and equipment, and bitter or unpalatable feeds. Amid all of his interest in science and technology, Catron also urged attention to the "tender loving care factor" as a stimulus to get hogs to work more eagerly.

Damon Catron issued these predictions at a time when he was taking stock of his own situation. In December 1959, Iowa State University asked him to submit a list of his activities over his fourteen-year career at the school-a list that included increasing the staff involved in swine research from two to forty, expanding the swine herd from thirteen head to 347, extending the value of the physical plant from $14,000 to nearly a million dollars, and acquiring over $539,000 in grant funds. In January 1960, Catron nonetheless announced his resignation from Iowa State and accepted a position with an Iowa feed manufacturing company.

Overall, this trajectory in Catron's career suggests that swine research had reached a level of stability and that one principal researcher could now reap in industrial employment what he had sowed during his period of academic research work. During the fifteen years that encompassed Catron's career in Ames, tremendous and rapid changes took place in the American swine industry.

The sights, sounds, and smells of the Midwestern landscape changed drastically as hogs moved from pasture and to increasingly confined operations in artificial environments. Increasing the pace of breeding, farrowing, weaning, rebreeding, and finishing had altered the pig's life cycle. The animal's physiology changed through the development of "meat-type" hogs engineered to reduce backfat measurements and feeds designed to get animals to market weight more quickly and efficiently.

Animals also became more dependent on pharmaceutical interventions, as farmers used antibiotics and other veterinary medicines as routine insurance against disease threats. Each of these changes altered the relationship between hog and the hog producer. The latter increasingly gained the ability to transfer responsibilities of labour and energy conversion onto the former, thus streamlining their operations along industrial lines.

Hog producers also saw their occupation grow increasingly connected to the emerging agribusiness complex that included commercial feed manufacturers, pharmaceutical firms, housing and equipment makers, government policymakers, and consumers. Hog existence became increasingly regulated through manipulations of their physiology, social activities, spatial environment, and connections with the natural environment.

A common thread impacted both hog and hog producer as both witnessed the disappearance of their independence and autonomy, a process similar to what had occurred in so many other industries. In addition to the calculus of costs, profits, and efficiencies, many other factors provided impetus for the changes in the hog's organism and environment.

After years of false starts, farmers and agribusiness leaders finally brought an industrial programme to a farm animal that seemed firmly bound to natural ecological rhythms and traditional market circumstances. On the farm, putting soybeans into rotation with corn made land too valuable for pasturing hogs, while manufactured nitrogenous fertilizers further reduced the incentive to graze livestock for their manure.

Increasing mechanisation made it possible for a single farmer to handle larger and more profitable acreages and proved an especially pertinent aid for the efficient handling of feed and manure.

Off the farm, various external pressures also induced farmers to reshape the hog to play a more vital role in the postwar economy and society. Increasing government manipulation of corn and hog prices, memories of the feed and agricultural labour shortages of World War II, popular pressure to apply wartime science to civilian ends, and changes in the behaviour of grocery store consumers who demanded leaner pork products all contributed to the redesign of these animals.

Such changes also had geopolitical implications that went beyond Corn Belt hog lots and farrowing stalls. In the words of one historian of American foodways, "the Cold War...was waged and perhaps even 'won' through kitchen debates, grain deals, breadlines, and the Green Revolution."

The cold war intensified Americans' interest in technological successes and, in the words of another scholar, "created the sense of an automation race" with Soviet bloc counterparts.

Americans justified technological innovations for their own sake and because increased production and bountiful supermarkets could be highlighted as symbolic of the free enterprise system. Along these lines, President Eisenhower's Agricultural Trade Development and Assistance Act and President John F. Kennedy's Food for Peace Programme both included policies that encouraged consumption of American-grown protein products and helping "friendly" nations with the meat and grain surpluses that remained.

It was no coincidence either that Soviet premier Nikita Khrushchev's 1959 tour of the United States included visits to both an American kitchen and to Iowa's efficient corn and livestock operations. For all of these reasons and more, the versions of the pig that had been adequate for centuries were no longer suited for the demands of the postwar world. Many actors participated in the process of creating the industrialized pig. By the 1950s, most of the players in the drama were already on the stage-pharmaceutical firms, feed manufacturers, animal husbandry experts, regulators, lobbyists, and more. Some farmers made an effort to slow the pace of these changes, and even in the late 1960s some 65 per cent of the hogs in Illinois were raised on some pasture. In all, though, little could stop trends towards antibiotic feeding, confinement, and other aspects of industrialized hog production.

The number of farms in Iowa that produced pigs fell from 181,400 in 1940 to 134,500 in 1960 and to 84,900 in 1970. The pace of consolidation intensified thereafter; by 2000, the number of pork producing farms in Iowa had dwindled to 10,500. Nowadays, empty and abandoned hog lots are a common sight in the rural Midwest. Even so, the story is not over.

SCIENTIFIC AND INDUSTRIAL RESOURCES FOR BIOTECHNOLOGY

As far as scientific and technological production in Latin America, university–industry research relations are almost non-existent. Universities in Latin America are more oriented to learning and teaching than to doing research.

One might say that the only practical relation between universities and industry is that of supplying the latter with a qualified labour force for professional and operative work. For the sake of illustration, let us look briefly at the case of Mexico, where university research is highly concentrated. The National Autonomous University of Mexico (UNAM) alone concentrates 50 per cent of all research funds in the country. Nevertheless, UNAM's research budget represented only 22 per cent of its global budget for 1988. The rest represented wages and other operating expenses.

The Mexican dilemma is that its science and technology system produces so little technology that most of it has to be imported. Still, there are four categories of university–industry links: training of professionals, creation of basic and applied knowledge, provision of technical services, and production of technology.

Part of the explanation for the weakness of university–industry links in Mexico is the near absence of a commercial criterion in research. Thus, the prevalent criterion tends to be defined in epistemological terms, which may be crucial for science but irrelevant for technology. For technology development, what matters is economic convenience, not epistemological relevance.

There are, however, some institutions doing cutting-edge biotechnology research in Mexico. In total, there are about 25 centres with biotechnology projects broadly defined. If we narrow the definition of biotechnology to the newer techniques, such as those based on recombinant DNA, the generation of hybridomas, cell and tissue culture, and protoplast fusion, then the number of institutions is reduced to only seven. The rest are involved in traditional biotechnology research.

Of those seven, three belong to the National Autonomous University of Mexico, two to the National Polytechnical Institute, one to the Autonomous Metropolitan University and another one to Chapingo Autonomous University. All of these universities are located around Mexico City, although three of the top research centres involved are decentralized in neighbouring cities (one in Irapuato and two in Cuernavaca). The pioneering institution in promoting industry links has been the National Autonomous University of Mexico (UNAM), reputed as the largest university in Latin America. It created a "Centre for Technological Innovation" in 1984 to market UNAM's science and technology. Although the explicit objective of the new centre is to market UNAM's science and technology, a former director declared to Expansión, the Mexican version of Fortune magazine, that they wish to preserve their identity as an academic institution.

They are not after tailor-made research, therefore, for the specific needs of companies. Rather, companies should find out what UNAM's researchers are doing that might be of interest to them. The director of one of the firms that has supported UNAM's research stressed that business firms should take advantage of Mexican technology, because it is cheaper and should allow them to compete better in international trade.

In sum, Mexico does have several excellent research groups working in the frontiers of biotechnology. But the number of researchers is very small in comparison to those in advanced capitalist countries, both in universities and in firms. In developing countries, there tends to be a very weak if not totally absent link between universities and industry. With such great differences in the quantity of scientists and the differing patterns of university–industry linkages, it is hard to be very optimistic about the possibilities for developing countries to compete successfully in such a dynamic field as biotechnology on the basis of their own resources.

It is necessary, therefore, to undertake a more realistic analysis regarding the actual scientific potential of developing countries and what public policies it would take to translate research into products that might solve their specific socioeconomic and environmental needs. Unless the proper policy incentives exist, it is unlikely that these needs will be taken into account by the private firms doing biotechnology research, or by scientists from advanced countries who have other concerns as their central motivations.

ECONOMIC SCALE AND BIOTECHNOLOGY

Contrary to what optimistic analysts of biotechnology hold, it seems that the main tendencies indicate a bias favouring the large scale, rather than scale neutrality.

That is to say, far from favouring small-scale farm operations, the new biotechnologies will tend to reinforce the bias introduced by modern agricultural technologies by requiring strong capital investments. With the new technologies then, the agricultures of developing countries are likely to require a smaller workforce, even if we assume that many of the products of biotechnology may be adopted by those countries.

The alternative might be that, if new technologies are not adopted, and given the global scope of today's world market, agriculture in developing countries may cease to be a viable economic activity altogether. This would be a major paradox, for about 60 per cent of the population in developing countries depends on agriculture. The following examples of new biotechnology products clearly reflect an economic bias towards large-scale production units. These are the implications of the "technological paradigm" represented by modern agriculture.

The generation and sale of cattle and sheep embryos is generally reserved for large and modern production units, which utilise the most advanced genetic technologies. Commercial cloning of livestock is becoming a reality. There are research groups in the U.S. that are developing cell fusion methods with the potential to create multiple, perhaps unlimited, copies of individual cow-and sheep-embryos.

The target in cloning is to produce genetically predictable animals, with superior milk-and meat-production capabilities. University of Wisconsin Professor Neal First and colleagues have successfully cloned cow embryos, but it will be years before they can determine the quality of the resulting breed. For what they clone is not the mother cow but the embryo itself, *i.e.*, an already fertilized egg. One of the goals of researchers is to improve the cloning technique to develop the entire process in test tubes and eliminate the need for embryos from the cows.

One implication of this type of research is that, in the future, it will be possible to produce "elite" animals for meat and milk production. This genetic improvement of animals, combined with the use of growth hormones, promises to revolutionise production in these fields, and they will probably displace large numbers of farmers, given the large-scale bias. The dictum for this type of modern operation will continue to be, perhaps in an exacerbated manner, "get big, or get out of farming."

Socially, this trend points to a future agriculture that will require not much more than a per cent or two of a country's labour force. The main concern will be with how to employ productively, with adequate incomes, those people who are displaced from agriculture. In the U.S., this stage has already been reached: only 2 per cent of the labour force is directly employed in farming.

Another factor that reinforces the previous interpretation regarding the economic bias towards the larger scale is that new agriculture will probably combine two or more products of the current technological revolution, including computers, for the integrated management of agricultural enterprises. It is very likely that a systems approach to management will be reinforced. For instance, and continuing with the example of livestock production, the most sophisticated dairy operations may use embryo transfer technologies to improve the herd, while using computers to maintain exact yield records per cow and to monitor their feed requirements.

The result will be that only large-scale operations will be able to make the heavy capital investments required for such integrated systems for managing resources. All of this reinforces the economic-scale tendency to discriminate against small-scale producers and accelerates the process of "depeasantisation without full proletarianisation".

The dilemma involved in this process is that, although farming expels many of its workers because of increased capital-intensity, the larger economy cannot absorb all those redundant workers. This is the source of the swelling "informal sector" in Latin American economies.

THE FUTURE OF INDUSTRIAL BIOTECHNOLOGY

In the USA in 2001, the Biomass R&D Board had surveyed 134 industrial sites that were manufacturing bioproducts and found 440 of which produced bioenergy. According to a report published in 2001 by McKinsey and Company, this consultancy reckoned that about 5 per cent of global chemical sales were derived in part form industrial biotechnology.

That figure could reach 20 per cent by 2010, if chemical firms, whose polymer innovation had stalled in recent decades, became convinced by the economics of bio-based products. It means that industrial biotechnology will be competing in a market worth $280 billion, of which McKinsey assumed that it might capture about $160 billion. The report by McKinsey and Company showed that the greatest impact would be on the fine chemicals segment, where up to 60 per cent of products could be derived from biotechnological processes by 2010.

Enzymes and fermentations were already used in the production of flavours and fragrances, while other markets would still be dominated by conventional chemistry through 2010 and beyond. The first applications in the largest volume segments – polymers and bulk chemicals – have already been commercialized.

However, in these largely cost-driven segments, a number of technological advances and policy measures would determine the ultimate uptake of industrial biotechnology.

McKinsey's estimates showed that the chemical industry alone could generate additional added value up of to □11 billion or □22 billion per annum by 2010, depending whether uptake was fast or slow.

Two inputs would contribute to this. One was lower costs for raw materials and processing, combined with smaller scale investments in the fermentation plants. The other was the additional revenues from innovative or performance-enhanced products.

As biotechnological processes become cheaper, the economic contribution of industrial biotechnology will increase. That is why DuPont, in cooperation with Genencor, was trying to decrease by 25 per cent the production costs and by 20 per cent the investment costs of its polyester Sorona, before launching its commercial production in 2004.

Among the incentives for industrialists to invest in bioproducts or biofuels, which were still more expensive than their chemical equivalents, the OECD suggested to include the real cost of the environmental impact of industrial processes in market prices.

The American Institute of Chemical Engineers proposed a total cost assessment method, which attempted to figure out in dollars all the risks posed on the environment, as well as on health and safety. Such a method and others like the 'green index' should entice an increasing number of enterprises to seize the opportunities offered by the bio-industry not only for the environment but also for the consumers and shareholders.

Burrill and Co., a San Francisco-based venture capital firm, has created a $50 million fund to invest in early-stage start-ups focused on biomaterials and bioprocessing. In the USA, representatives from different governmental bodies, industry, agriculture and academia worked together on a project called 'Vision 2020' with a view to boosting industrial biotechnology over the next decades backed by low feedstock prices and the support of the agricultural community in the USA.

The Farm Bill, passed in May 2002 by the US House of Representatives, authorized $5 million in 2002 and $14 million a year from 2003 through 2007 to fund biomass research and award grants to build 'biorefineries' – factories converting biomass into chemicals, fuels and energy. The Bill also requested the government to give preference to purchasing bio-based products.

Meanwhile, the European Bioindustries Association has proposed to the European Commission to form a strategic alliance to design a so-called Technology Platform. In order to stimulate global companies to invest in industrial biotechnology and in Europe, bridging incentives, more effective regulatory processes and low-cost feedstock were key factors. Another switch that may prove crucial to watch is the move from ethanol to biotechnologically produced hydrogen. The driving power would not come from internal combustion engines but from fuel cells using hydrogen.

In this regard, genomics may be the answer. Using the money Celera Genomics had raised, Craig Venter, who designed the whole-genome shotgun sequencing for both the human and microbial genomes, set up the Institute for Biological Energy Alternatives (Rockville, Maryland), with a view to studying microorganisms that produce hydrogen.

One example is that of Carboxydothermus, discovered in a hydrothermal vent (deep-sea volcanic spring) off the coast of Russia, which draws its energy from reacting carbon monoxide with water, and produces hydrogen as a waste product.

Another of Craig Venter's projects was to create a bacterium with a completely synthetic genome, leaving out the genes for carbohydrate synthesis that normally use hydrogen ions, so that the bacterium could devote all its energy pathways to producing hydrogen. Such unicellular micro-organism would not be able to live outside the laboratory because it would lack the biochemical mechanisms to survive there.

Discovering new materials for the industry while drawing on natural resources also speaks well for the prospects of industrial biotechnology. For instance, Nexia Biotechnologies, based in Quebec, is using technology similar to that of GTC Biotherapeutics to manufacture spider silk in goat's milk. Spider silk has been produced for the past 400 million years and has been adapted to a wide range of uses. This is an opportunity human beings could seize and use biotechnology-derived spider silk in many applications, from surgery to the design of new textiles.

BIOMINING AND BLEACHING OF ORES

Chile is the world's first-biggest producer of copper. While in 1990 the production target had been fixed at 2.5 million tons for the year 2000, this figure was superseded in 1995, and the production exceeded 5 million tons in the late 1990s. It has been exploiting copper since the 19th century with the help of local and foreign entrepreneurs, mainly from the USA. In 1971, copper mines were nationalized during the brief socialist episode of Chile's president Salvador Allende.

Since then and despite the liberal economic policies conducted under the dictatorship of Augusto Pinochet and thereafter since 1990 when Chile returned to democracy, there has been no privatisation in the mining sector with 80 per cent of the mines still presently owned by the state. Because of the importance of copper mining and production in Chile's economy, the country has often been nicknamed the 'Saudi Arabia of copper'. Copper's physico-chemical properties and multiple uses, as well as a strong demand for the product from Asia (mainly

China) have resulted in a high rise of international prices during the first half of 2004.

Codelco, the National Copper Corporation, created in 1976, controls 20 per cent of the world's copper reserves and in the late 1990s produced around 2 million tons of the metal, accounting for some 13 per cent of world production. Copper represented 40 per cent of Chilean exports and one-third of Chile's earnings. According to its director-general, Codelco is planning to raise its annual production up to 3 million tons in 2012.

To illustrate the fast development of copper mining and production in Chile, one could cite the small city of El Salvador, located at an altitude of 2,600 metres and 1,000 km north-west of the capital Santiago; it was created during the 1950s with foreign partners in the middle of a desert area. Nowadays, out of its 10,000 inhabitants, 2,000 are working in copper mining and production.

Codelco bears the expenses of housing costs of its workers as well as of the education of their worker's children. Water recycled from the copper mills is being used for producing vegetables, fruit and flowers, grown in greenhouses, as well as for feeding small livestock.

The ore extracted from El Salvador's mines and containing about 2 per cent of copper is treated in a village in the same region, Llanta, so as to obtain a concentrate with 40 per cent of copper. This concentrate is thereafter treated in the furnaces of Potrerillos (99.5 per cent of copper) and then catalysed to produce plates of pure copper (99.9 per cent). El Salvador is one of the five main divisions of Codelco.

Its annual production only reached 85,000 tons of copper in 2003, but there are many prospects for extension. For instance, El Salvador could follow the model of Chuquicamata – the biggest open-air copper mine in the world, with an annual production of 600,000 tons – which was created at the beginning of the 20th century, partly with funding from US billionaires, including the Guggenheim.

Chuquicamata extends over a huge basin 5-km long, 3-km wide and more than 850 m deep. The copper ore is extracted in the middle of the basin and transported in huge trucks to the surface to be treated. Every day, at 5 pm, a new explosion renews the stock of raw material to be processed.

At El Teniente, south of El Salvador and another division of Codelco, a gigantic subterranean mine is being dug from the top of a mountain down to the surface level.

Explosions are carried out in the heights; rocks are being ground downward, and loaded on a train at surface level. Codelco is not excluding any innovative solution to better exploit the 'national treasure'.

In its first foray into the world of international corporate takeovers, Codelco and the Canadian company Noranda announced by mid-August 2000 a $1.76-billion bid for Rio Algom, representing a 'major breakthrough' in Codelco's efforts to expand. But the joint bid was quickly topped by Billiton of the United Kingdom, with a $2-billion agreed offer. Noranda and Codelco had to rethink their strategy, maybe with a new bid.

Codelco's expansion plans were boosted in 2000 by a recovery in copper prices from a 12-year low in May 1999, from 60 cents to about 85 cents a pound. It also benefited from a cost-cutting programme forced on it by low copper price and cash costs were down 7 per cent to 41.1 cents a pound in the first half of 2000. Codelco then partnered with Phelps Dodge of the USA, the world's second-biggest copper producer, to build the El Abra mine in Chile as well as prospecting alliance with the Mexican Grupo Penoles, the world's third-biggest producer. It also had a joint venture with Billiton, called Alliance Copper, which pioneered the commercial extraction of copper from low-grade ore using bioleaching.

While the Chilean government ruled out selling Codelco, which gave up 10 per cent of its annual revenues to fund military expenses, the debate about privatisation was continuing, particularly as large-scale utility sell-offs initiated in the late 1980s were nearing completion. The company nevertheless, seemed to be steadily moving towards a private-sector model.

Bioleaching had been used only for the recovery of copper from wastes until the mid 1980's. The process was upgraded thanks to multidisciplinary and multi-institutional research efforts by several universities, research institutes and the productive sector, with the support of the Chilean government, the United Nations Development Programme (UNDP) and the United Nations Industrial Development Organisation (UNIDO). In 2000, there were over five mines using bacterial bioleaching, and there were several projects to expand the use of this technology in the future, so as to raise its contribution to national copper production over 8 per cent to 10 per cent.

The 5 million tons of copper produced in the late 1990s were the result of processing 400 million tons of ore, including 40 million tons treated via bioleaching. Back in Chile, the country had been exploring ways and means for applying bioleaching to gold production. Following the duplication of gold production in the country during the 1990s and the prospects for a 50 per cent increase in production during the three-year period 2000-2002, the position of Chile is being fostered in this area.

In July 2001, Chile's ministry of economy, the National Committee for Science and Technology Research (CONICYT) and the National Development Corporation (CORFO), in the presence of Chile's president, R. Lagos, entrusted the company BioSigma S.A. with pursuing a leadership in biotechnology applied to copper mining.

This attracted a Japanese corporation, Nippon Mining and Metals Co. Ltd., to join BioSigma S.A., to establish leadership in biotechnologies involved in copper mining and in about one year the new company was created. Codelco's contribution was 66.6 per cent of the company's equity, while that of Nippon Mining and Metals Co. Ltd. amounted to one-third of equity. The Japanese corporation's contribution was not only financial, but it also opened the doors of for advanced technology and industry transfers among the countries involved.

BioSigma S.A. was the first example of an alliance between two major mining groups and a government which allocates competitive funds to attract research groups from Chilean and foreign universities to help industry transform basic knowledge into innovative products and services.

Consequently, BioSigma S.A. elaborated a plan of technological development, which included the identification of specific microorganisms, technologies for the production of biomass of these microorganisms as well as the identification of specific genes encoding proteins that facilitate the bioleaching of copper sulfurated ores.

It was estimated that a minimum of one thousand genes would be identified in every microorganism. The expected outcome of the plan included technologies aimed at improving the current biomining processes, as well as more advanced solutions relating to the cloning or designing bacteria that increase bioleaching efficiency.

BioSigma S.A.'s business plan set up short, medium and long-term tasks, and it was foreseen that BioSigma S.A. would recoup earnings from technology sale in 2005. A major challenge for copper mining in Chile is the high concentration of arsenium and chloride in copper ores as well as the necessity to keep the mining wastes on the site and not to remove and dump them near urban settlements, *e.g.* around the city of Calama in the extreme north of the country.

It was therefore necessary to design a safer and cheaper bioleaching process by arsenium-resistant and chloride-tolerant microorganisms, to be applied not only to new mining sites, but also to mining wastes from which the maximum amount of residual copper would be extracted. Nippon Mining and Metals Co. Ltd., which is also a shareholder of several copper mines in Chile, including the well-known huge mine named La Escondida, has built a $150-million pilot plant, in association with Codelco, to test the applicability of the bioleaching process named BioCop and developed in South Africa by the BHP consortium; in this pilot plant, tests included the use of chloride-tolerant hyperthermophilic micro-organisms.

The University of Antofagasta which has set up a Centre of Biotechnology and Molecular Biology dealing with industrial and environmental biotechnology, has signed a biomining programme research contract with BioSigma S.A. Codelco, Nippon Mining and CONICYT have contributed $2 million, $2 million and $1 million, respectively, to the biomining programme, which is part of CONICYT's Genomics Chile – the national genomics programme devoted to mining on the one hand and to agriculture and food on the other.

The allocated budget allowed the building of BioSigma S.A.'s new reference laboratory in Santiago, the maintenance of the company's head office also in Santiago and the funding of four research projects – three Chilean and one Japanese. Three of the four projects were funded for the three-year period 2004-2006, and the fourth one was an innovative, exploratory research project to be carried out by the University of Antofagasta. The latter aimed at developing innovative bioremediation treatment of toxic mine effluents, *i.e.* transforming toxic metallic compounds into harmless substances. The three other research projects focused on sequencing the genomes of bioleaching microbes, cloning them to improve the bioleaching process, and on bioinformatics.

4

Immobilization

The use of plants as production systems for vaccine antigens has been actively investigated over the last 15 years. The original research focused on the value of this expression system for oral delivery based on the hypothesis that plant-expressed antigens would be more stable within the digestive tract and would allow for the use of the oral route of administration to stimulate a mucosal immune response. However, while first conceived for utility via the oral route, plant-made antigens have also been studied as classical immunogens delivered via a needle to model animal systems.

Antigens have been expressed in a number of whole plant and cell culture systems. Several alternative expression platforms have been developed to increase expression of antigens or to elicit preferred immunological responses. The biotechnological advances in plant expression and the immunological testing of these antigens will be reviewed in this paper focusing primarily on diseases of livestock and companion animals.

PLANT-MADE VACCINES

Advances in plant biotechnology have opened the door for the use of transgenic plants as production systems for vaccine antigens. It was over a decade ago when it was first demonstrated that an antigen could be expressed in plants and fed as a vaccine to produce a measurable immune response. Since then, the expression of vaccine antigens in planta has progressed to allow for the expression of bacterial, viral and self-antigens in transgenic whole plants and plant cells. Many authors have discussed the value of this expression system which includes decreased manufacturing costs and the potential for alternate delivery routes including oral delivery. Much of the focus of plant-made vaccine studies has been directed towards their utility for oral delivery. The primary advantages of this delivery route are the increased likelihood of eliciting a mucosal immune response and the ease of vaccine administration.

The latter is considered key for disease control in third world countries (humans) and for decreased stress and ease of administration in animals. While oral delivery is attractive, the amount of protein required has been estimated to be as large as 1 mg. This is only a target since the actual amount will depend on antigen presentation, immunogenicity and formulation of the antigen. Nevertheless, the amount of protein required will likely be large for oral delivery and, subsequently, represents a challenge for many plant expression platforms. Vaccines targeted for use by subcutaneous administration will likely require significantly less antigen.

Despite research efforts in this area of vaccinology, to date, there are no commercial products, human or veterinary, that have been approved by the regulatory authorities who license these products, although Dow AgroSciences has entered the regulatory path for licensure of a plant-made vaccine for Newcastle disease in poultry.

In this review, we focus on the current status of in planta expression including whole plant and plant cells and also on the practical application of this technology beyond laboratory phase testing.

METHODOLOGIES FOR EXPRESSION OF VACCINE ANTIGENS IN PLANTA

Expression Platforms

Initial studies of plant-made vaccines utilized easily transformed and manipulatable model crops such as tobacco. Subsequent work has focused on crop plants, most targeted for human use, but some which could overlap for use in animal vaccines including potato, tomato and corn. A considerable amount of work has been done using potato tubers as a platform. The development of an ethanol inducible promoter that allows inducible expression of a transgene in harvested tubers is an interesting improvement that would enhance the utility of tubers as a delivery system for recombinant proteins.

Transgenic, non-cooked tubers have been shown to deliver amounts of antigen necessary to provide some level of protection in disease models. Problems reported have been the slow degradation of protein in harvested and stored tubers and the fact that raw potato is not palatable for humans and has triggered gastrointestinal side effects in volunteer subjects. These and other reasons have led to the more recent shift in opinion that some degree of processing may be necessary or preferred for many of the plant-expressed materials.

A plant species more thoroughly tested for expression of proteins has been corn. Production of proteins in corn seed forms a basis of several studies including those reported by the company Prodigene. The advantages of corn are that the transgenic protein in the dry seed tends to be stable and the seed can be easily stored and transported. In addition, corn crops can be rapidly expanded, depending on demand. Importantly, the matrix of the corn seed extracts has been shown to provide protection against degradation of protein during passage through the gut.

This latter point is of particular significance to this and other orally delivered plant-made materials. The challenge is to thoroughly investigate stability of the expressed protein in the range of gut conditions present among individuals in an animal population and even at different times within an individual. Transgenic corn has been used to obtain protective immune responses against Escherichia coli heat-labile enterotoxin (LT) and an antigen from the swine transmissible gastroenteritis (TGE) virus (TGEV) and LT-B, the B subunit of LT. Tomato has also been a popular plant species for expressing vaccine candidates, providing both containment and ease of processing. The fruit has the advantage over some other expression platforms in that it can be consumed fresh, eliminating the need for post harvest processing. Moreover, existing technologies are available for processing the fruits, formulating with adjuvants and delivering the vaccine. Plant species that are targeted specifically for animal health and have been used for vaccine production include clover, alfalfa and pigeon pea.

The latter plant is used as a possible source of livestock feed in developing countries. Included among livestock-related antigens expressed in alfalfa are those from foot and mouth disease, TGEV, classical swine fever virus and Mannheimia haemolytica. An alfalfa expression system is the basis of the commercial enterprise Medicago, Inc., which has developed alfalfa transformation and protein production strategies to make it an attractive alternative to more common platforms.

In particular, Medicago's work has focused on identifying promoter elements that provide high levels of expression using constitutively expressed promoters or inducible promoters. Inducible promoters will be of value for expressing proteins that are detrimental to plant growth and development. Although most researchers developing plant-made vaccines use field-grown plants, some work has been done with cell cultures.

One of the advantages cited for field-grown plants relative to cell cultures grown in bioreactors is the lower cost of goods. However, cell-culture-based platforms do not have to address the issues of containment, manufacturing process controls and chain of custody. Often, the cost of regulatory approval has not been included in the final cost estimates of field-grown products when the above arguments are made.

The most widely available cell culture used for testing plant-made vaccine product concepts is the tobacco line BY-2 (or its sibling culture NT-1). This culture was developed as a low-nicotine line that rapidly increases cell mass. Although used extensively for expressing transgenic proteins, its use for producing vaccines has been largely limited to work recently reported. Several antigens have been expressed in tobacco suspension cultures by these groups including those from avian influenza virus (AIV), Newcastle disease virus (NDV), E. coli heat-labile toxin and infectious bursal disease virus (IBDV). All transgenic cultures were shown to produce proteins that retained their biological activity by either in vitro tests or direct disease challenge trials.

Another plant species tested for expression of recombinant proteins is safflower. This specialized technology (StratosomeTM, SemBioSys Genetics, Inc) incorporates a unique expression and purification strategy in which the protein of interest is fused to oleosin, the protein that coats oil bodies in safflower seeds. After the seeds are crushed, the oil bodies can be floated away from other seed materials by centrifugation, resulting in a highly purified preparation of transgenic proteins.

Gene Constructs

Most reports in the literature describe early stage product development in which putative protective antigens are expressed in plants and the resultant transgenic material tested for its ability to either elicit an immune response in an animal, or preferably, to provide protection against disease. The majority of these research programmes has been led by academic researchers with little or no consideration of intellectual property ownership and, hence, often use promoters, genetic elements responsible for expressing genes, that are not in the public domain.

The most common among these is the cauliflower mosaic virus 35 S promoter or its variants. This promoter is generally considered a high expressing dicot promoter. Some researchers have used variants of the Agrobacterium mas gene promoter for expression of vaccine candidate proteins because of their reported high expression. The patatin gene promoter was shown to express higher in potato tubers than different versions of the 35 S promoter and the sunflower polyubiquitin gene promoter. Optimization of promoters for expression of vaccinerelated proteins in plants is an area that needs further exploration.

Accumulation of transgenic protein depends not only on the transcription rate of its gene, determined principally by the promoter used, but also on the steady state level of its mRNA, efficiency and rate of translation of the messenger RNA (mRNA) and stability of the protein. Expression of transgenes in heterologous systems has demonstrated the importance of altering codon usage to increase the rate of protein production. The impact of host cell codon bias depends on how closely it matches the codon bias of the source organism of the transgene.

In some cases, altering the codon composition of a transgene is not important, but in others, it can lead to dramatic increases in expression. High mRNA levels require correct and efficient processing of the mRNA, including transcription termination, polyadenylation and splicing. Plants vary from other eukaryotes mechanistically in how they process RNAs and, further, monocots can differ from dicots. Typically, plants tend to be less precise in the sequences that drive these processes. For example, other eukaryotes use a single predominant sequence for directing RNA polyadenylation, whereas plants have many. The challenge for using plants to express antigens is to completely understand the nuances of the plant gene expression machinery and design transgenes accordingly so that signals are not misinterpreted by plant expression machinery.

Most studies testing potential transgene vaccine candidates for animal health ignored plant-specific gene expression issues. How this has impacted expression levels of those proteins, which tend to be low, is worth exploring for the more promising protective antigens. Other gene elements that can affect expression levels are translational enhancing elements that occur in the 50-untranslated region of mRNAs. Most widely incorporated into constructs is the omega element from tobacco mosaic virus (TMV) gene, which increases translation by recruitment of eIF4F translation factor. This element improved expression 11-fold when it was used to express a hepatitis B antigen. Targeting of proteins to different cellular compartments can be critical to the success of potential vaccines expressed in plants. In many cases, targeting the protein to the endoplasmic reticulum (ER) increases accumulation relative to expression in the cytoplasm.

There are exceptions to this, which may be due to glycosylation interfering with protein folding or stability. Proposed reasons for the improved protein accumulation in ER include the presence of high levels of chaperones (proteins assisting with protein folding), a reduced level of proteases, a more oxidizing environment and presence of protein disulfide isomerase, an enzyme which forms disulfide bonds in proteins. Targeting to the ER requires the presence of a native or heterologous signal peptide on the N-terminus of the protein.

This directs the protein to the endomembrane system, but retention in the ER requires the addition of a retention signal (*e.g.*, KDEL, SEKDEL and HDEL) to the C-terminus of the protein. Addition of these sequences to proteins normally not targeted to the ER may alter protein conformation, a key consideration when designing expression strategies. Absence of the ER retention sequence on a protein steers it into the default secretory pathway, which leads to secretion of proteins that are below the size retained by the cell wall or insertion into the plasma membrane for proteins with membrane anchoring regions.

Transgenic Proteins in Plants

Accumulation and conformation of transgenic proteins are dependent on correct or aberrant post-translational modifications. Most post-translational modifications found in non-plant eukaryotes have also been demonstrated to occur in plants, including glycosylation, S-acylation, N-myristoylation, phosphorylation, and acetylation. One of the most common post-translational modifications is glycosylation, with plant proteins containing both N-linked and O-linked glycosylation. However, plant glycosylation patterns can be significantly different than those found in mammals. This may be important for in vivo use of plant-produced antibodies for passive immunization or other uses. Among the most noted differences is the presence of b(1,2)-xylose and a(1,3)-fucose residues in plant glycans that are not found in mammals. These different substitutions may lead to rapid clearing of antibodies from the bloodstream when administered to animals.

As proteins are transported into the ER, high mannose glycans are added to accessible asparagines in the amino acid sequence context Asn-Xaa-Ser/Thr and more complex glycans are built on the mannose framework in the Golgi apparatus, including addition of xylose and fucose at particular sites in the glycan. Plants can be genetically modified to produce glycans that are more similar to those found in animals.

Expression levels of transgenic proteins are generally from undetectable to 0.4% of total soluble protein (TSP) and, hence, are considered low. While there can be differences between plant species in the level of expression of particular proteins, this difference tends not to be large. For edible vaccines, the expression level typically obtained is often sufficient to elicit some immunological response by feeding, although whether the level would be commercially viable is not reported. Nevertheless, expression from transgenic inserts located in nuclear genomes and driven by highly expressed promoters clearly will not always be sufficient without some purification or concentration of the antigen. Some alternative production systems for plant-made antigens include oil bodies, viruses and plastids.

Plastids (the most commonly used plastid in plant transformation being the chloroplast) contain a genome that is separate from the nuclear genome. Importantly, there can be multiple copies of the genome per plastid and multiple plastids per cell, resulting in the presence of up to 10,000 copies per cell. This large gene copy number has led to extraordinarily high levels of expression, with typical increases of expression being from 500-to 4000-fold relative to nuclear-located transgenes.

The potential for producing large amounts of commercially important proteins led to the development of the commercial enterprise Chlorogen. An important advantage of plastid transformation is that the transgenic insert is targeted to specific regions of the chloroplast genome by sequences that are homologous to chloroplast DNA that flank the transgenes of interest. Hence, issues of position effects and gene silencing that plague nuclear transformation experiments are not problems with plastid transformation experiments. Key to the success of this expression platform is that chloroplasts have the machinery to fold and process most proteins correctly. Several plants have been transformed using this technology, but transformation of others of potential interest has not been reported.

Daniell *et al.* argue that carrot has advantages for oral delivery since carrots can be easily stored, eaten uncooked and, importantly, the plant is biennial; hence, preventing spread of seed and pollen if harvested after the first generation.

A challenge with this technology can be to ensure a high and stable level of plastids carrying the transgene in the crop plant of interest. The primary limitation of the technology is the lack of protein glycosylation. While the activity of many proteins is not affected by glycosylation, it can be critical to protein conformation or function in other cases. The use of this technology for development of vaccines is illustrated by expression of a tetanus toxin gene and a canine parvovirus peptide fused to cholera toxin B subunit (CTB) at 25 and 31%, respectively, of TSP in tobacco.

How antigens are presented to the immune system can drive the type of immune response that occurs. Peptides or protein subdomains can be fused to other proteins or protein domains that stabilize the protein and improve the accumulation of the protein. The most widely reported protein fusion partners are non-toxic variants of bacterial exotoxins such as heat labile endotoxin (LT) from E. and cholera toxin (CT) from Vitrio cholerae that bind to particular gangliosides on mucosal surfaces. Oral administration of LT or CT fusions target the proteins to M cells in Peyer's patches and subsequent delivery to antigen-presenting cells within this tissue elicits a mucosal response, which can be essential for protection against many enteric pathogens. This has been a widely used strategy for experimental vaccine development. It is not known how an immune response to LT or CT *per se* might alter responsiveness to the target antigen.

A chimeric protein consisting of a protein of interest and a viral coat protein configured in such a way that the coat proteins can assemble into a virus or virus-like particle (VLP) elicits a strong immune response. Key to the success of this approach is that VLPs may be more readily taken up by antigen-presenting dentritic cells.

A limitation of this strategy is the requirement to minimize disruption of critical coat protein structure, which is done, for example, by insertion of peptides into an external loop of the coat protein. Cowpea mosaic virus, tomato bushy stunt virus, TMV and alfalfa mosaic virus (AlMV) have been used successfully to express high levels of peptide sequences in plants. An example of this approach was the demonstration of disease protection against virulent challenge and IgG and IgA responses in mice against rabies by oral administration of AlMV coat protein fused to a peptide containing antigenic determinants from the rabies virus. There has also been an extensive use of plant virus expression platform in HIV vaccine development.

Immunogenicity of Plant-expressed Antigens

Several viral subunit proteins, expressed in plant systems, have been tested for their potential as oral vaccines. These include mainly viral antigens from pathogens such as Norwalk virus, TGEV, foot and mouth disease virus (FMDV), rotavirus and parvovirus. Not surprising, there are only limited studies involving bacterial pathogens probably due to the complex nature of bacterial pathogenesis and the inability to clearly delineate proteins that are critically associated with protection. While some of these have been characterized immunologically, few studies conclusively elucidate the apparent unique properties of plant-made vaccines or the immune response elicited by using them.

One of the strongest supporting rationales for expression of vaccine antigens in planta is associated with the hypothesis that when delivered in plant material, vaccine antigens will be protected from potential degradation in the gut, leading to increased vaccine efficacy. It has been suggested that plant antigens are 'bioencapsulated' within the plant cell walls and cell membranes and therefore have increased resistance to digestion. This hypothesis was supported in an experiment comparing the immunogenicity of VP2 antigen of canine parvovirus expressed in tobacco.

Mice were gavaged with either soluble plant extracts of VP2 antigen of canine parvovirus at two different doses (200 or 1000 mg) or with tobacco leaf extract (600 mg) and the data demonstrated that numerically higher serum IgG titers were obtained in the mice that were gavaged with leaf extract. The authors attributed this increased response to the protective effect of the plant matrix to digestion.

However, since tobacco leaves were unpalatable to mice, gavage was used as the method of delivery of the antigen. It is not known if direct delivery to the stomach limits the potential for antigen presentation along the upper gastrointestinal tract.

The ability of orally administered plant-expressed vaccines to stimulate a mucosal response was reported by Wigdorovitz *et al*. Using a mouse model of bovine rotavirus (BRV), these researchers investigated whether females vaccinated orally would transfer lactogenic immunity to their offspring. In this study, the eBRV4 epitope of the VP4 protein of BRV was expressed in transgenic alfalfa and either extracted from the alfalfa for intraperitoneal injection in mice or fed as an oral vaccine to mice.

Both routes of administration stimulated serum antibody titers as determined by ELISA. Female mice receiving these vaccines were bred and 5 days after birth, offspring from these females were challenged orally with a virulent strain of BRV. Offspring of vaccinated females had significantly less clinical disease 24 h post challenge compared to their non-vaccinated cohorts.

The authors asserted that this vaccination regime stimulated lactogenic immunity; however, no data were presented to support this conclusion.

Colostrum was not assessed for specific antibody. More importantly, the hemendothelial placentation of mice allows for transfer of large amounts of immunoglobulin across the placenta. Therefore, the protection seen in the offspring of the vaccinated dams may be associated with transplacental transfer of protective IgG and not necessarily lactogenic transfer of rotavirus antibodies as the authors surmised.

Vaccine Testing in Target Animal Species

Although much has been written about the potential for production of vaccines against diseases of livestock, there is scarce information addressing immunogenicity or protective efficacy using relevant challenge models in the species of interest. Studies in mice or similar model species can show that the plant-expressed antigen retains the ability to drive an immune response but extrapolation to the target species, especially with respect to efficacy, must be done cautiously since model infections in mice frequently have divergent pathogenesis, and by extension, alternative protective immune responses, even when disease in the model appears to physically resemble that in the natural host.

There are as yet no published reports of vaccination and challenge studies in ruminants despite activity by several research groups. Dus Santos and Wigdorovitz (2005) recently reviewed work examining three pathogens of particular significance for the cattle industry of Argentina, FMDV, BRV and bovine viral diarrhea virus (BVDV) expressed in various plants including Arabidopsis thaliana, alfalfa and potato. Vaccines against the FMDV and BRV have demonstrated immunogenicity after intraperitoneal (FMDV and BRV) or oral (BRV) exposure and resistance to challenge with virulent virus in mice.

Researchers in India have Targeted the Highly Contagious and Deadly Disease

Rinderpest, still prevalent in parts of Africa, the Middle East and South Asia, where lack of facilities and extreme heat constrain maintenance of the cold chain necessary to retain viability of the current modified live vaccine. They expressed the protective viral hemagglutinin antigen in pigeon pea (Cajanus cajun), the leaves of which are a common fodder crop readily consumed by livestock, with the intent of producing an easily stored and transported edible vaccine. In North America, pneumonic pasteurellosis remains an economically important disease of feedlot cattle.

Several protective antigens of the principal causative bacterium M. haemolytica have been identified and researchers have succeeded in expressing the critically important leukotoxin (Lkt) in clover and alfalfa, for production of an edible, needle-free vaccine that will be easy to administer, reducing stress on the animals, which can itself precipitate disease, and avoid costly trim losses at injection sites. The majority of plant-made vaccines have been tested in model animal systems. Vaccination and challenge trials that have been reported in target species are rabbit hemorrhagic disease virus (RHDV) in rabbits; TGEV in pigs; mink enteritis virus in mink; canine parvovirus in dogs; and IBDV, AIV and NDV in chickens.

Potato plants expressing VP60 protein of RHDV have been shown to stimulate a protective immune response against challenge when administered four doses (at monthly intervals) of potato leaf extract containing 12 mg of recombinant VP60 with Complete (day 0) and Incomplete (days 30, 60 and 90) Fruend's adjuvant. Vaccination was by the subcutaneous route for the first vaccination, and intramuscular for the subsequent vaccinations. Rabbits were challenged with a virulent strain of RHDV and vaccinated animals were completely protected from disease while nonvaccinated control rabbits died. Although these data strongly support that this vaccine stimulated a protective response in rabbits, limited animal numbers (n=4) did not allow statistical validation.

Efforts in swine seem to be focused exclusively on development of an oral vaccine for TGE. Although current injectable vaccines are reasonably efficacious in passive protection of piglets by immunization of sows, they do not induce production of mucosal antibodies and the need for multiple injections poses a management hardship.

The TGE spike protein (S) which drives a virus neutralizing response was successfully expressed in tobacco plants (Nicotiana tabacam) and three injections of piglets with leaf tissue suspensions induced antibodies in serum detected by ELISA that were weakly virus neutralizing. Streatfield *et al.* were able to demonstrate that in swine, plant-made vaccines can stimulate protective immunity. Ten-day-old piglets, fed with seed corn expressing the spike (S) protein of TGEV, were protected against a virulent challenge with TGEV. The protection was similar to that induced by a commercial modified live vaccine (MLV) with none of the risks associated with use of an MLV. The authors noted that the duration of symptoms was shorter in pigs receiving the modified live vaccine.

This may be in part due to differences in the breadth of the immune response stimulated by the modified live vaccine as compared to the non-adjuvanted subunit corn seed vaccine. Since there are limited data generated to demonstrate that plant-made vaccines are protective in host species, these data are a very important advance towards the development of this type of vaccine technology. More recently, Lamphear *et al.* reported an increase in antiviral antibodies in serum, colostrum and milk of sows fed with a corn-based oral TEGV vaccine following priming by a single injection with conventional vaccine. No attempt to assess passive transfer of protection to piglets was reported. There have been relatively more plant-made vaccine vaccination and challenge studies completed in poultry. Vaccination and challenge trials are reported for IBDV and NDV.

For IBDV, live attenuated vaccines are currently employed for control, but highly attenuated strains give only moderate protection, while less attenuated vaccines can produce disease in vaccinates. Wu *et al.* expressed the VP2 protein of IBD virus in A. thaliana and demonstrated a serum antibody response in chickens fed with leaf extracts of the transgenic plants. Subsequently, the plant-made vaccine was compared to a commercial vaccine of intermediate attenuation given by ocular/nasal drop for protection of specific pathogen-free broiler chickens against challenge infection at 4 weeks of age. The commercial vaccine was administered at 1 and 3 weeks of age, while the plant-made vaccine was given either orally by gavage, as five doses 3 days apart commencing at 1 and 3 weeks, or injected subcutaneously in Freund's complete adjuvant at 1 and 3 weeks. One group of birds received the commercial vaccine at week 1 as a priming dose and the oral vaccine regime at 3 weeks to boost. Data were pooled from three independent replicates of the experiment, each with ten chickens per group. Chickens in all vaccinated groups produced antibodies by 1 week after the first vaccination that increased following the second dose. Protection against droplet challenge with virulent virus varied from 90% in the orally vaccinated group to 60% in the groups receiving the plant material subcutaneously.

The commercial vaccine alone or with the oral plant vaccine boost provided approximately 80% efficacy. All control chickens vaccinated orally with saline or a non-transformed plant extract succumbed to challenge. These studies are encouraging with respect to the potential to induce protective immunity by feeding transgenic plant material; however, as these researchers acknowledge, current practice in the industry is vaccination in ovo, often using robotics, and further work is needed to determine the relative ease of vaccination and efficacy of oral plant-made vaccines.

Protection against IBDV challenge was also demonstrated by Miller *et al.*. These researchers expressed the VP2 protein of IBDV in tobacco cell cultures and produced vaccines using this antigen for demonstration of efficacy by challenge. In the study, birds received three doses of vaccine (days 7, 21 and 35) either via the subcutaneous route of administration, via the oral gavage route of administration (non-adjuvanted) or via the intrabursal route of administration (non-adjuvanted).

All vaccines delivered via the subcutaneous route of administration were protective against challenge except for the VP2 antigen formulated with drakeoil emulsion adjuvant. Those vaccines delivered via oral gavage or intrabursal were not protective. Further application of the plant-made vaccine technology in poultry has been demonstrated by Cardineau *et al.* In these studies, the hemagglutinin/neuraminidase protein (HN) of NDV was expressed in tobacco cell cultures and chickens were vaccinated with plant-cell-produced HN (adjuvanted or non-adjuvanted) at 2 days of age and 16 days of age via the subcutaneous, intranasal, oral gavage or oral gavage followed by feeding routes of administration. Birds were challenged 14 days after the second vaccination. One hundred per cent of the non-vaccinated control animals died of NDV and 100% of the positive control birds (receiving native NDV subcutaneously) survived. Protection was evident in the birds receiving the plant-cell-produced HN vaccine by the subcutaneous route of administration (non-adjuvanted, 75%; and adjuvanted, 80%), but no protection was evident (100% mortality) by the intranasal, oral gavage or oral feeding routes of administration.

These researchers also reported results from a study where birds received either 20 mg of plant-cell-produced HN protein or 250 mg of plant-cell-produced HN protein with or without MPL–TDM adjuvant (Corixa, Inc). While 100% of the unvaccinated controls died post challenge, 80% of the birds receiving either 20 mg of HN adjuvanted with MPL-TDM or 250 mg of HN non-adjuvanted were protected. Additionally, 100% of the birds receiving adjuvanted 250 mg of HN were protected and 60% of the birds in the non-adjuvanted 20 mg group were protected. The authors concluded that this plant-cell-produced vaccine provided excellent immunity when administered subcutaneously, however, since these antigens are non-invasive and do not have a natural affinity for mucosal surfaces, immunity was not generated by mucosal administration.

Although not of primary veterinary interest, it is worth noting some experiments that have been completed using plant-made vaccines where humans are the target species. Edible plant-made vaccines are of particular interest for protection against enteric diseases since antigen delivered to the intestine has the greatest potential for stimulating relevant mucosal immunity. Human volunteers ingesting three doses of transgenic potatoes expressing the LT-B gene encoding the E. coli heat-labile enterotoxin (average 0.75 mg dose71) showed antibody-secreting cells in the circulation 7–10 days post ingestion and 10 of 11 had 4-fold rises in serum of LT-specific IgG; six of these also had increased antigen-specific IgA. Similarly, 19 of 20 volunteers who ingested two or three doses of potatoes expressing Norwalk virus capsid protein (NVCP) developed increases in the number of circulating antibody-secreting cells and four developed antigen-specific IgA in feces. The importance of hepatitis B infection globally has spurred considerable interest in development of plant-made vaccines for effective oral delivery in regions where vaccination by injection is problematic.

Efforts over the past several years were recently reviewed by Streatfield. Transgenic lettuce (Lactuca sativa) expressing the surface antigen of hepatitis B virus (HbsAg) induced serum IgG responses in humans fed twice at a 2 month interval. Serum anti-HbsAg antibodies also increased in 10 of 16 humans who ate three doses of transgenic potatoes and in 9 of 17 who ate two doses. Edible plant-made vaccines have also been investigated as a format for vaccination of adults against measles.

Challenges for the Technology

At the Plant Made Pharmaceutical Conference held in Montreal, Canada in 2005, the potential limitation of plant-made vaccines was discussed. While the focus was on orally delivered vaccines targeted for human health, many of the same conclusions are similar for animal health vaccines. There are clear potential benefits of oral delivery of plant-made vaccines; however, there are still questions of expression levels and efficacy in general.

For example, will adjuvants be required to obtain the desired immune response or will VLPs be a major platform due to their expanded immune regulatory potential ? Regulatory approval for a plant-made vaccine has not yet been achieved and, for some production systems, remains uncharted. In fact, for field-grown materials, multiple requirements will be demanded, both for efficacy of the products and proper stewardship and containment of the plants. To our knowledge, no animal health product derived from field-produced plants has been advanced significantly towards product licensure. Additionally, dependent on the type of system utilized to express the vaccine antigen, the cost of registration and the potential cost of containment can have a significant impact on the commercial viability of the end product. One issue of oral delivery of transgenic proteins that is commonly discussed is the potential for oral tolerance. In fact, oral delivery of auto-antigens may lead to oral tolerance and reversal or reduction of autoimmunity. These reports involve animal models and have yet to be fully evaluated in the target host.

In the research that has focused on oral delivery of plant-expressed antigens as vaccines, there have been no examples of oral tolerance reported. The difference in biological outcome of oral administration of transgenic proteins may be dependent on the specific regimen used for the treatment. For full adoption of a vaccine dependent on oral delivery, optimization of vaccine delivery and a more thorough understanding of the immune response to vaccination will be necessary.

Since the first published account of the potential of plants as an expression vehicle for antigens, research has focused on antigen expression in plants and models of immunogenicity. Only very limited studies have investigated the efficacy of these plant-made vaccines in the animal host (human or non-human) and the impact of differences in the commonly used model mouse system versus the host animal are yet to be fully understood.

Although many different types of plants have been utilized and many different types of antigens (viral, bacterial and self-antigens) have been expressed, the full applicability of this technology platform for the production of vaccines for animal health will not be fully demonstrated until products are licensed for use in species of importance to veterinary medicine. The future of plant-made vaccines lies in the researcher's ability to develop vaccines against diseases of importance and to test these vaccines for their efficacy in the host. Advantages of the plant expression system over conventional systems will continue to be elucidated as experimental data amass in this new and growing field of vaccinology.

NEW HEALTH PRODUCTS FOR FARM ANIMALS

Animal diseases cost U.S. agriculture $17 billion each year, according to the Congress Office of Technology Assessment. To help prevent these losses, animal scientists are using biotechnology to develop an array of products to diagnose, treat, and prevent disease in farm animals. An area where animal biotechnology has already had a profound impact is diagnostic testing. Many animal diseases are difficult to diagnose. A veterinarian often has to wait hours or days for laboratory results to confirm a diagnosis. In the meantime, the veterinarian must either withhold treatment or risk using an inappropriate therapy.

Using biotechnology, scientists are developing fast, accurate diagnostic tests for many of the most common farm animal diseases. In some cases, a veterinarian can conduct the diagnostic tests right on the farm and immediately start the best therapy. Many of these new tests use monoclonal antibodies. Antibodies are proteins an animal's immune system produces in response to invasion by bacteria, viruses, or parasites. Each kind of antibody binds specifically with a characteristic part of a particular kind of invader. Scientists make monoclonal antibodies by fusing two kinds of cells. One is an immune system cell that produces antibodies, which bind to part of a particular disease-causing microbe; the other is a cancer cell.

The cell that results from the fusion inherits from the immune system cell the ability to produce antibodies, and from the cancer cell the ability to reproduce indefinitely.

This new hybrid cell, or hybridoma, acts as an antibody factory. The antibodies are called monoclonal because they are produced by identical "clones" of the original hybridoma-all will bind with the identifying structure on the surface of the microbe in question. One application of monoclonal antibodies is as a test for brucellosis in cattle. This bacterial disease often causes cows to abort pregnancies, and it can be spread to farmers and people who drink milk from infected cows. Vaccines protect pregnant animals from abortion, but vaccinated cows may become carriers of the disease. Conventional diagnostic tests cannot distinguish between the disease-causing microbe and the vaccine, which is made from the microbe. Thus, these tests cannot help the farmer determine which animals are carriers of the disease and which are not. But diagnostic tests using monoclonal antibodies are so specific, they distinguish between cattle that carry the disease-causing bacterium and those that carry only the vaccine. With the aid of these new tests, farmers can isolate carriers to prevent the spread of infection. Diagnostic tests are not just for diseases.

A monoclonal antibody test to detect pregnancy in animals is so simple that a farmer can read the test right on the farm. Monoclonal antibodies are also used in a test to detect a disease called scours in piglets and calves. Scours is a bacterial disease of newborn animals that causes diarrhea. Many of the afflicted animals die of dehydration. By providing quick, accurate diagnosis, the monoclonal antibody test for scours permits immediate administration of therapy.

The therapy for scours in calves is also based on monoclonal antibodies. Infected calves are fed monoclonal antibodies that coat the offending bacteria, preventing the microbes from causing diarrhea.

Some other therapies for animal diseases are based on recombinant DNA technology. Genetic engineering allows scientists to insert one or more new genes into an animal, plant, or microbial cell. In developing treatments for an animal disease, for instance, an animal gene is inserted into a microbe or an animal cell that can be grown in a laboratory culture. The inserted gene is one that produces a natural disease-fighting protein. The microbes or animal cells become minute factories, producing large quantities of the therapeutic protein. For example, genetic engineering is used to produce interferons and interleukin-2, natural proteins that fight viruses.

Veterinarians long have used antibiotics against bacterial diseases, but they have had few drugs to treat viral infections. Interferons and interleukin-2 not only kill viruses directly, they also stimulate the animal's immune system, improving the immune system's ability to fight disease naturally. One target of research regarding interferons and interleukin-2 is shipping fever, a disease of cattle caused by several viruses infecting an animal simultaneously.

Although the animal's immune system normally keeps the viruses in check, outbreaks of the disease may result from stress during transport to feedlots or market. Shipping fever costs the beef industry more than $250 million a year. Injections of interferons or interleukin-2 before shipping may help to augment the cattle's natural immunity and prevent shipping fever.

BETTER VACCINES TO PREVENT DISEASE

In addition to producing new diagnostic tests and therapeutic proteins, animal scientists are using biotechnology to develop vaccines to prevent disease. A safe, effective vaccine for swine pseudorabies, a fatal herpes-virus, is already in use. An advantage of the swine pseudorabies vaccine, and other vaccines made through recombinant DNA technology, is that they use only a small portion of the original microbe. Conventional vaccines use killed or weakened forms of the disease-causing microbe.

Sometimes, when the microbes are not completely killed or sufficiently weakened, a conventional vaccine may cause the disease it was supposed to prevent. Disease-causing genes are not included in the genetically engineered vaccines. Therefore, the recombinant vaccines build up the body's immunity without the risk of causing disease. Another advantage of recombinant vaccines is the speed with which they can be developed.

"Conventional vaccine development can take 20 or 30 years, maybe even 100 years," says Karen Jacobsen, D.V.M., University of Georgia College of Veterinary Medicine. "This is one reason why we don't have vaccines for a lot of animal diseases today. " Recombinant vaccines are being developed for foot-and-mouth disease, a highly contagious viral disease that infects cattle, sheep, and other animals.

Although this disease has been eradicated in North America, it still causes substantial livestock productivity losses elsewhere, particularly in developing countries. Infected herds must be slaughtered, and contaminated ranches must be left idle for months to prevent new outbreaks of the disease. Conventional foot-and-mouth vaccines are made by weakening the virus that causes the disease. These vaccines sometimes revert to the virulent state, and they have caused outbreaks of the disease in Europe. The foot-and-mouth vaccine made with biotechnology cannot cause the disease because, as in other biotech vaccines, the disease-causing genes have been removed.

The biotech vaccine does not need to be kept cold and, therefore, can be used in developing countries. In addition to being used in treating swine pseudorabies and foot-and-mouth disease, recombinant DNA technology is being used to develop vaccines against dozens of livestock and poultry diseases, such as scours and a variety of respiratory disorders. Some vaccines may be injected into eggs so that chickens are immunized before they hatch. Most vaccines protect against viral or bacterial infections, but using genetic engineering, researchers in New Zealand and Australia have developed a vaccine against a parasitic disease. Farmers must destroy the meat from sheep infested with a species of tapeworm common in those two countries. The new vaccine, which protects sheep from the tapeworm, may save farmers millions of dollars annually.

PROTEINS FOR ANIMAL NUTRITION

Natural proteins called somatotropins, or growth hormones, help animals convert feed into muscle or milk. Using recombinant DNA technology, scientists have been able to develop bacteria that produce commercial quantities of somatotropins.

They have found that administering small additional amounts of these proteins to animals helps to increase the animals' efficiency of feed conversion, thereby saving the farmer money. Porcine somatotropin (PST) may help improve human health, as well as lower the farmer's cost of production. PST improves feed efficiency in hogs by 15 to 20 per cent. It also reduces fat deposition, allowing PST-treated hogs to provide consumers with leaner cuts of pork.

The ability to produce lean pork has enormous implications for improving human health by reducing dietary fat and cholesterol. Another growth protein, bovine somatotropin (BST), is being administered to make dairy cows more efficient, thereby lowering the dairy farmer's costs. Regular administration of small amounts of BST increases a cow's milk yield by 10 to 25 per cent. Although feed intake also increases, this feed increase is proportionately less than the increase in milk production.

Studies indicate that BST can improve the milk-to-feed ratio by 5 to 15 per cent. Extensive laboratory testing has shown that the meat and milk of treated animals contain no more PST or BST than the meat or milk from untreated animals. Other tests have shown that these proteins do not act on the human body, and therefore, they are safe to eat. Nevertheless, the use of BST remains controversial.

Testing is under way to make sure that animals are not adversely affected by treatment with somatotropins. Another protein that can be administered to animals to increase feed efficiency is called growth hormone releasing factor (GHRF). While not a growth protein itself, GHRF causes the animal to increase production of growth proteins.

IMPROVED ANIMAL BREEDS

Long before anyone knew about genes or DNA, people were selectively breeding animals to produce desirable traits, such as strength, feed efficiency, and disease resistance. Selective breeding is a crude technology, because each cross between two animals mixes hundreds of thousands of genes, both desirable and undesirable. Recombinant DNA technology is now making it possible for scientists to breed animals with great precision. One or more new genes can be inserted into an animal embryo without disturbing the rest of the animal's genes.

The resulting animal with the new gene or genes is referred to as a transgenic animal. Despite the differences in technique, the scientist using recombinant DNA has the same objectives as the selective breeder: improving efficiency, disease resistance, and other desirable traits. But recombinant DNA provides a wider array of possible new traits, since recombinant techniques allow animal scientists to insert genes not only from the same or closely related species, but also from distantly related animals, plants, and even microbes. One possible application of gene transfer in animals is to develop dairy cows that produce more nutritious milk. Bottle feeding human infants with cow's milk is not an ideal substitute for breast feeding, because cow's milk differs from human milk in some important components. Scientists are looking for ways to transfer genes into dairy cows that will enable the cows to produce milk that more closely resembles human milk.

Advances in Control of Parasitic Diseases Continue to Rely on Biotechnology Parasitic disease continues to be a major health problem worldwide in humans and domestic animals. Applying biotechnology has led to progress in prevention and control in many areas, particularly in regard to diagnosis. Parasites, like most organisms, carry repetitive sequences in their genomes. These are sometimes tandem repeats (back-to-back in a chromosome), or they may be located at random throughout the genome. Often, repetitive sequences are intergenic and noncoding, and are thus far more subject to evolution and change, increasing the likelihood that they will be genus or even species specific. Since detection of these sequences is not dependent upon gene expression, which may vary over time, repetitive sequence-based assays allow detection of the parasite at any life stage.

The effectiveness of repetitive sequence-based polymerase chain reaction (PCR) methods of identification has been demonstrated for parasites of several clinically important genera (1). A PCR assay has also been developed to differentiate two closely related nematode species that are virtually indistinguishable by morphologic examination (2), as well as for diagnosis of Toxoplasma and Cryptosporidium infections (3). Attempts to develop anti-parasite vaccines have met with multiple failures. The problems often result from lack of post-translational modification by currently used vector-host systems and the inability to regenerate the secondary and tertiary structure of the native immunogens. Antigen presentation in recombinant parasite vaccines has also been thought to be a problem resulting in failed efficacy.

There is now some evidence to indicate that cloning of genes for antigens of interest to yield chimeric molecules expressed in the BCG strain of Mycobacterium bovis may provide improved antigen presentation to stimulate both humoural and cellular responses (4); however, it is unlikely that future applications of this technology will be simple. Nonetheless, work is proceeding on the application of biotechnology to the development of vaccines against parasites in most food animals. A particularly striking exception to the marginal successes reported in other systems has been a recombinant vaccine for a parasitic helminth of sheep (5). This product yielded greater than 94% protection in experimental work and is now available as a commercial product. A further possibility is the application of anti-idiotypic monoclonal antibodies. When monoclonal antibodies are prepared against a monoclonal antibody, it is possible to obtain, in a very small number of cases, an antibody which mimics the original antigen. Results of work with Cryptosporidium perfringens epsilon toxin suggest that animals immunized with anti-idiotype monoclonals can be protected against toxin challenge.

BIOTECHNOLOGY AND ANIMAL HEALTH

The potential areas of impact for biotechnology on the control of infectious diseases and improved animal health are manifold. They include the identification of susceptible stock, tests for the rapid and early diagnosis of disease, an understanding of the mechanisms of pathogenesis and the development of safe and effective vaccines or prophylactics. The principal methodologies which will enable these targets to be approached are those of monoclonal-antibody production and recombinant-DNA technology. The benefits of selecting stock on the basis of resistance to specific infectious agents would appear at present to be restricted to those areas of the world where the importance of a single disease predominates over all others (*e.g.* trypanosomiasis).

Whilst the possibility of selecting on an antigen non-specific basis is likely to be of more general relevance, at present the development of vaccination strategies that can effectively control infectious agents is likely to be of greatest significance for improved animal health.

To this end monoclonal antibodies can be used in the development of diagnostic sera and reagents, and the characterization of virulence and adhesion determinants. Once identified such determinants can be isolated in a pure form from crude mixtures with monoclonal immunosorbents. Infectious diseases may be controlled by the prophylactic use of monoclonal antibodies, but this direct effect will be of limited use in animal production, by comparison with their potential use as 'educators' of the immune system in anti-idiotype vaccines. Recombinant-DNA technology can provide a range of potential immunogens including attenuated or non-pathogenic strains and subunit vaccines. Since the majority of infectious agents are met at mucosal surfaces, the problem that remains is how these vaccines may be delivered so as to provide effective presentation and so stimulate a protective response.

Mucosal Defence

The level of challenge to, and the degree of sterility maintained differs markedly between the various mucosal surfaces; from the gastrointestinal and upper respiratory tracts to the lower respiratory tract, reproductive tract and mammary gland which at least in health remain sterile. They are presented with a vast array of different antigens ranging from potential pathogens to relatively harmless dietary components. Together with the skin they provide the interface between the host and its environment. However, unlike the skin there are other physiological roles (*e.g.* gaseous exchange, absorption of nutrients, etc.) which restrict their ability to keep antigens out. Since the mucosal surfaces provide such a primary point of contact it is at these sites that the products of biotechnology might be expected to provide the most relevant and beneficial effects. For convenience the major part of the present review will be restricted to antigens presented to the gut. The immune system of the gastrointestinal tract is presented with two types of antigens, those capable of causing infection and those, such as dietary antigens, which in themselves impose no direct threat to the animal's health. The type of response stimulated is critical, for whilst failure to respond to potential pathogens may be fatal, an over-reaction to dietary antigens may lead to damaging hypersensitivity responses. Latter may in turn lead to altered handling of potential pathogens.

Therefore, although the most obvious area of impact of biotechnology will be directly on infectious agents, regulating responses to food antigens may also be of significance.

The Response to Food Antigens

Three responses have been described that may protect an individual from damaging hypersensitivity reactions to dietary antigens. Immune exclusion mediated by locally produced IgA may reduce the absorption of intact proteins across the gut mucosae and in certain species it has been shown that any antigen that is absorbed may be eliminated via bile. These processes which restrict the access of antigenic material to the immune system are not absolute and the development of an acquired specific hyporesponsive state is probably of greater significance.

Thus orally induced tolerance may actively down-regulate the response to dietary proteins. Before the induction of this 'protected state' all animals pass through a transient phase of hypersensitivity, during which time damaging allergic reactions can be demonstrated.

In farm species it has been shown that following weaning, piglets are sensitive to antigens in the post-weaning diet and since these changes occur before the proliferation of *Escherichia coli* it has been postulated that response to food proteins may be an important predisposing factor in the aetiology of post-weaning diarrhoea. The importance of gut flora and bacterial lipopolysaccharide in regulating the response to dietary antigens has been clearly shown in rodents, thus by the use of engineered strains of bacteria it may be possible to regulate the response to dietary proteins and so prevent the transient allergic responses that reduce the gut's ability to handle potential pathogens.

This brief description of the regulation of the response to food proteins attempts to serve two purposes. First, it may indicate a role for engineered strains in the controls of hypersensitivity reactions and the consequences thereof. Second, the type of response generated to food proteins differs markedly from that required to eliminate infectious agents. It is then perhaps not surprising, as will be discussed later, that dead mucosal microbial vaccines are generally not very effective.

The Response to Infection

The implications of biotechnology on animal health directly related to infections are numerous; it is therefore proposed to discuss them sequentially from diagnosis to vaccine development. The conventional methods for the identification of the causal agent have in the past depended heavily on an ability to isolate and grow the organism in pure culture. The use of monoclonal antibodies should aid both the identification of infectious agents and their virulence factors. Unlike conventional anti-serums, which by virtue of the nature of the immune response are very heterogenous, monoclonal proteins have a unique specificity which allows such discrimination to be made.

For example, by use of the enzyme-linked immunosorbent assay (ELISA) it is possible to detect subtle antigen differences between strains of an organism as well as to characterize the immune response of the infected animal. Once having identified a particular virus, for example as being important in the pathogenesis of diarrhoea, how then might one proceed towards an effective vaccine? Consider for instance that virus X has been shown to be an important pathogen but is closely related to similar commensals. It can produce disease, and animals that recover from infection show solid protection. In neutralization tests it has been possible to show the specificity of X, but it is very difficult to grow! Further progress may then involve the development of a sensitive immunoassay for research and sero-epidemiology. Production of a monoclonal antibody (either conventionally from rodents or preferably from infected farm species) may allow isolation of a strain-specific antigen. Subsequently the antigen should be produced by cloning and expressing the gene in, for example E. *coli,* by recombinant-DNA technology. Such a procedure may form the basis for the development of a vaccine.

Control of Infection

Without attempting to consider ways of eradicating an organism from the environment there are two types of approach to the preventionxontrol of infection within an animal. The first is essentially passive and involves little or no requirement of the host's immune system, whilst the second requires the stimulation of an active vigourous immune response.

Passive Protection

Monoclonal antibodies have been used therapeutically to passively protect neonatal pigs and calves against challenge with a lethal dose of a K99 positive pathogenic strain of E. *coli.* It has been pointed out that such an approach has a theoretical advantage over conventional vaccination of the dam, in that it does not require anticipation of the problem. Further, in this case where the disease is restricted to the first 24 h of life and the effects of the bacteria are contained within the intestine, it is directly treated by oral administration of the monoclonal antibody. A similar approach has been used to protect mice from intranasal infection with Respiratory Syncital Virus, but generally it would appear that the therapeutic use of monoclonal antibodies will be of benefit in specific, defined individual areas, notably in human medicine, rather than on a herd basis in veterinary medicine.

The rationale for this argument is the amount of antibody required, particularly if it is necessary that it is given orally. Monoclonal antibodies to infectious agents may, however, be of benefit in the 'education' of the immune system and this active process will be discussed later. Another passive form of treatment involves the use of anti-microbials, and in this context biotechnology has already made a significant impact. Human leucocyte interferon has been isolated by monoclonal anti-immunosorbents and on a commercial level recombinant interferon and interleukin 2 are now generally available.

Active Protection

Before considering the potential impact of biotechnology on the development of safe, effective vaccines, it is necessary to consider what is required of such products in order that they may stimulate a protective response. As has already been mentioned, the vast majority of infectious agents are met at a mucosal surface and therefore vaccines should be directed towards stimulating protection at these sites. To this end it is clear that by far the most effective method of eliciting such a response involves presentation of the antigen at the mucosal surface. In general parenteral vaccines have not been effective, and where exceptions exist they almost always involve some mucosal component (*e.g.* oral + parenteral).

The other exception to this principle involves immunization of the dam to achieve passive protection of the sucking neonate, by virtue of the presence of antibodies in colostrum.

In the case of domestic species, antibodies in colostrum are derived from serum and these may be effectively stimulated by parenteral immunization. Unlike colostrum, milk antibodies are locally produced and in this case, like other true mucosal secretions, mucosal immunization is required to stimulate production of antibodies.

Having established that mucosal vaccines are those of choice for both theoretical and practical reasons, it is, however, disappointing that it is surprisingly difficult to stimulate a vigourous, protective response via this route. As has already been pointed out, this is of benefit with regard to the 'harmless antigens' (*e.g.* dietary antigens) also presented to the mucosae. To date the majority of effective mucosal vaccines are living attenuated organisms reflecting the now generally held axiom that either replication or at least adherence are essential prerequisites of mucosal immunogenicity. Whilst there is now clear evidence of memory in the secretory immune system, vital if any vaccination regimen is to be effective, it has also been shown that the response is highly dose-dependent.

Thus where dead oral vaccines have been used, very large numbers of bacteria are required and even then the level of immunity is sub-optimal by comparison with that following the use of live organisms. Another problem associated with mucosal immunization, unique to the gastrointestinal tract, relates to an ability to pass through the stomach in a form that remains antigenic. In this context the use of capsules or liposomes may be of benefit. Alternatively the observed immunological link between mucosal sites, such that oral immunization leads to the appearance of antibodies in milk, may be of benefit.

The New Vaccines

Vaccines derived as products of biotechnology may take a variety of forms and rather than discussing them on a specific disease basis it is proposed to briefly describe the general feature of each type. The genes coding for protein or peptide antigens derived from a number of bacteria and viruses have been cloned and expressed in appropriate bacteria. For example, with the foot and mouth disease virus one of the structural capsid proteins, WI, has been identified as carrying the antigenic determinants necessary to raise neutralizing antibodies in experimental animals.

The viral RNA encoding for WI has been copied into double-stranded cDNA, inserted in a bacterial vector and expressed in *E. coli*. Subsequently a range of oligopeptides from VPI have been isolated or synthesized and a major antigenic determinant comprising nineteen amino acids identified. The question that remains is how these proteins or peptides may be presented to the immune system in order to stimulate a protective response. The isolated VPI protein, for instance, is not as antigenic as the virus and a 100 to 1000-fold excess on a molar basis is required in order to obtain satisfactory protection. With the oligopeptide it is necessary to couple to a carrier protein (*e.g.*, KLH) and although a single injection of 200 pg has been shown to protect guinea-pigs from virus challenge, to date no satisfactory method has been found to stimulate a protective response in cattle. Similar results have been obtained with *E. coli* antigens (K88ab, ac, ad and K99 have all been cloned and expressed) but it still remains to be determined how a protective mucosal response may be stimulated.

Whilst to date this synthetic peptide vaccine approach has been disappointing, its theoretical attraction over conventional dead vaccines remains since it avoids the requirement for handling large volumes of infectious materials, or vaccinal strains that grow poorly in bulk-suspension cultures, and the risk associated with failure to completely inactivate the virus before incorporation into a vaccine. An alternative approach involves the use of attenuated strains and these may take a variety of forms, and have been used against a number of agents. Deletion mutants of Aujeszky's disease virus have been prepared and although shown to have greatly reduced virulence are still highly immunogenic in pigs.

Similarly an attenuated strain of *Salmonella typhi,* Ty2la, which is unable to utilize galactose, has been derived. The organism is able to survive for a limited number of divisions, but antigenicity remains unaltered. For another enteric pathogen, cholera, an appropriate mutant is also available. In this case it produces only the B subunit of the toxin and thus, whilst being able to bind to the epithelial cell and give good protection, does not generally produce diarrhoea. Where attenuated strains of a particular organism are not available or appropriate the use of live recombinant vaccine may be of benefit.

Intranasal inoculation with recombinant *Vaccinia* virus has been shown to protect mice against upper and lower respiratory tract infections with influenza. Alternative vectors to *Vaccinia* are also available, for example yeasts or engineered non-pathogenic strains of E. *coli* that produce only the B subunit of the labile toxin, and so bind to enterocytes but do not produce diarrhoea. The general acceptability of live attenuated vaccines remains to be determined and the possibility of developing techniques whereby peptide vaccines can be delivered in a way that stimulates a protective mucosal response remains attractive. In this context the use of peptide conjugates coupled to molecules such as cholera toxin or lectins that bind avidly to epithelial cells may be of benefit.

Alternatively the potential role of antiidiotypic antibodies (monoclonal) may be of interest, it having been recently shown that such antibodies may prime for protection against E. *coli*. The capacity of biotechnology to produce a range of antigenic moieties derived from infectious agents would appear to be unlimited. Further, it should not be difficult to identify those antigens that are important in the pathogenesis of disease.

The major problem that remains is how best these products might be presented to the immune system in order to stimulate a protective immune response. It is therefore to this target of understanding how antigens are presented to and processed by the mucosal immune system that intense future research should be directed.

APPLICATIONS AND IMPLICATIONS OF ANIMAL BIOTECHNOLOGY

In most developing countries, biotechnological applications relating to livestock need to be suitable for animal owners who are resource-poor small-scale operators who own little or no land and few animals. ivestock is becoming increasingly important to economic growth in developing countries and the application of biotechnology is largely dictated by commercial considerations and socio-economic goals. Using technology to support livestock production is an integral part of viable agriculture in multi-enterprise systems. Livestock are part of a fragile ecosystem and a rich source of animal biodiversity, as local species and breeds possess genes and traits of excellence.

Molecular markers are increasingly being used to identify and select the particular genes that lead to these desirable traits and it is now possible to select superior germ plasm and disseminate it using artificial insemination, embryo transfer and other assisted reproductive technologies. These technologies have been used in the genetic improvement of livestock, particularly in cattle and buffaloes, and the economic returns are significant.

However, morbidity and mortality among animals produced using assisted reproductive technologies lead to high economic losses, so the principal application of animal biotechnology at present is in the production of cheap and dependable diagnostic kits and vaccines.

Several obstacles limit the application of biotechnology at present: there is a lack of infrastructure and insufficient manpower, so funding is needed if resource-poor farmers are to benefit from biotechnology.

The developing world is grossly unprepared for the new technological and economic opportunities, challenges and risks that lie on the horizon. Although it is hoped that biotechnology will improve the life of every person in the world and allow more sustainable living, crucial decisions may be dictated by commercial considerations and the socioeconomic goals that society considers to be the most important. Globally, livestock production is growing faster than any other sector, and by 2020 livestock is predicted to become the most important agricultural sector in terms of added value. The use of biotechnology will lead to a distinct shift in the economic returns from livestock.

Livestock production currently accounts for about 43% of the gross value of agricultural production. In developed countries livestock accounts for more than half of agricultural production, while in developing countries the share is about one-third. This latter share, however, is rising quickly because of rapid increases in livestock production resulting from population growth, urbanisation, changes in lifestyles and dietary habits and increasing disposable incomes.

THE LIVESTOCK ECONOMY IN DEVELOPING COUNTRIES

Livestock is becoming increasingly important in the growth of agriculture in developing economies. The contributions made by livestock to both agriculture and gross domestic product (GDP) have risen, at a time when the contribution of agriculture to GDP has fallen. The demand for livestock products is a function of income, and sustained growth in per capita income, rising urban populations and changes in diet and lifestyle are fuelling growth in livestock production. Livestock production contributes to socioeconomic development in many ways, by augmenting income and employment and reducing the incidence of rural poverty. Though the role of livestock in ensuring nutritional security is recognised in mixed crop-livestock systems, the importance of livestock goes beyond direct food production. Livestock supply draught power and organic manure to the crop sector, and hides, skins, bones, blood and fibre are used in many industries.

Thus, livestock are an important source of income and employment, helping to alleviate poverty and smooth the income distribution among small landholders and the landless, who constitute the bulk of the rural population and the majority of livestock owners.

In addition, livestock can easily be converted into cash and thus act as a cushion against crop failure, particularly in less favoured environments. By enabling crop residues and by-products to be used as fodder, livestock production contributes positively to the environment.

Animal owners in the developing world are predominantly resource-poor small-scale operators with little or no land and few animals, who must operate within the constraints of the local climate and who have limited purchasing power and little access to resources or opportunity to determine resource allocation for animal production.

The situation of the poorest livestock owners is fast deteriorating owing to the fragmentation of limited holdings, exhaustion of land resources and increasing human and animal population pressure.

Low livestock productivity in many developing countries is considered to reflect, among other things, the inadequate supply of animal husbandry and veterinary services. Veterinary services have traditionally been provided by the State, but financial constraints have limited the availability and effectiveness of public services.

THE IMPLICATIONS OF TECHNOLOGY

A major benefit of agricultural research and technology is that the purchasing power of the poor increases, because both average incomes and access to staple food products are improved. Studies by economists have provided empirical support for the proposition that growth in the livestock sector affects the whole economy. Rapid growth of livestock production has stimulated demand for and increased the value of land, labour and nonagricultural goods and services, thereby leading to overall economic growth. The poor spend a relatively high proportion of any additional income on food, so increases in livestock production achieved through the use of biotechnology can have major nutritional implications, particularly if the technology is aimed at the poorest producers.

However, studies have revealed that the commerci-alisation of agriculture has reduced the nutritional security of the poor. Once production of milk, meat or eggs has been enhanced through the use of technology, it is hoped that it will also make a significant difference in other areas such as nutrition, prevention of diseases, healthcare and other management practices. It is in these areas that biotechnology shows promise and is currently being used. Green Revolution technologies (*i.e.* those technologies designed to improve the efficiency of agricultural processes and increase crop productivity by relying on the extensive use of chemical fertilizers/pesticides and heavy machinery) are intended to be used in package form (*e.g.* new plant varieties supplied with recommendations on fertilizer, pesticide and herbicide rates and water control measures); however, among livestock producers many components of these technologies have been taken up in a piecemeal, often stepwise, manner. The sequence of adoption is determined by availability and by the potential cost savings.

The sequential adoptions of crop management technologies for rice and wheat have been assessed in detail, but few similar studies have focussed on livestock production in developing countries. Evidence from the People's Republic of China, Mexico, South Africa and India suggests that small farmers have had no more difficulty than larger farmers in adopting the new technologies. The question, therefore, is not whether biotechnology can benefit small-scale resource-poor farmers, but rather how biotechnology can address the agricultural problems faced by farmers in developing countries.

Biotechnology is a promising new tool in the development of applied agricultural technologies. The challenge is to focus this potential on the problems experienced by developing countries. The introduction of multi-enterprise systems or, more broadly, agricultural diversification is seen as the way forward for agriculture in the developing world; such systems could lift small-scale and marginal farmers out of poverty.

For example, rotating rice and wheat cropping with dairy farming yields higher profits. Introducing multi-enterprise systems involving livestock enhances the purchasing power of farmers and helps them to obtain nutritional security. It also generates rural (both farming and non-farming) employment, thus preventing excessive migration to urban areas, which is a common problem in developing economies. Multi-enterprise systems also support the natural environment and contribute to capital formation, thus leading to higher overall growth in the agricultural economy. The technology, infrastructure and institutions now exist to make the application of biotechnology in the context of a multi-enterprise system involving livestock production economically viable.

Global Advantage from Livestock of Developing Countries

The multiplicity of genes, species, populations and agroecosystems in the developing countries of South and South East Asia, the People's Republic of China, Africa and Latin America is viewed as a valuable resource for the genetic improvement of livestock on a global scale. The livestock in these countries, which are an integral part of a fragile ecosystem, are a rich source of animal biodiversity. Buffaloes, sheep, goats, camels and zebu cattle have adapted to their regional environments over thousands of years and have provided an important source of sustenance for the population of the region.

Livestock production in the developing world has a number of advantages over production in more developed countries, for example:

- The unique and valuable production traits of buffaloes, cattle, sheep, goats and camels
- The low-input production system
- The low unit cost of production
- The lean meat produced from sheep, goats and buffaloes
- The considerable biodiversity
- Animal breeds that are resistant to stress and to particular diseases
- The ability of the animals to survive on high-roughage feeds
- The potential for biopharmaceutical developments to lead to significant benefits
- The potential for expanding the microbial food, feed and leather industries
- The integrated production system tailored to the local ecology
- The potential for integrating knowledge and industry.

Several genes and desirable traits have been identified in the livestock of developing countries in Asia and Africa, and some of the livestock species and breeds from these countries have become major contributors to the economy of South America.

Examples of breeds from the developing world that are particularly important on a global level are:

- Buffaloes that produce milk with a high fat content or with the protein quality required to produce mozzarella cheese
- Goats from cold dry regions that produce pashmina and toos (the finest wool in the world)
- Black bengal goats that carry a gene for high prolificacy
- Garole sheep that carry genes for twinning
- Andaman goats that are highly tolerant of salt
- The yak and mithun that are adapted to high altitude
- The camels, sheep and goats that are adapted to a tropical arid environment and can tolerate feed with a high lignin content
- The many species that are resistant to stress or to particular diseases.

Economic Impact of Technologies

The genetic resources possessed by animals in developing countries often affect economic development. Traxler has discussed the economic impacts of biotechnological innovations, but the research and policy options need separate consideration. Animal biotechnology is the result of a multistage process, involving research, development, testing and registration, production and marketing. The goal is to develop a technology, process or product that has clear commercial potential and can be commercialised after due testing and regulatory approval. Developing countries find it difficult to develop biotechnology because the facilities or resources needed to complete all of the stages in the process are often lacking. However, several technologies from developed countries have been successfully adopted by developing countries.

The impact of technology can be analysed by estimating the growth of total factor productivity (TFP) in livestock production. Not many TFP studies on livestock have been reported.

However, separate TFP estimates for the aggregate crop and livestock sectors have been made. TFP analysis has shown a shift towards larger, more commercial and more intensive production systems and has further revealed that, as specialisation has developed over the past decade, the importance of backyard livestock production has declined and the importance of specialised household and commercial enterprise has increased. Studies from India have shown that technological input is responsible for about 45% of total output growth and that the TFP growth may be as much as 1.8%.

TECHNOLOGIES THAT HAVE A SPECIFIC IMPACT IN DEVELOPING COUNTRIES

There are a large number of technologies that have been developed for or adapted to the livestock of both developed and developing countries. However, the major technologies that are used effectively in livestock production in the developing world include conserving animal genetic resources, augmenting reproduction, embryo transfer (ET) and related technologies, diagnosing disease and controlling and improving nutrient availability.

TRANSGENICS

Although gene-based technologies have the potential to improve the efficiency of livestock production, thereby ensuring better returns for the farmers, the economic impact of transgenics in the livestock sector will be much less than in the crop sector. However, the global adoption of genetically modified (GM) crops, which were grown on 67.7 million hectares in 2003 compared with 2.8 million hectares in 1996, has had a substantial impact on livestock feed. It is estimated that the United States of America (USA), Argentina, Canada, Brazil and the People's Republic of China have 63%, 21%, 6%, 4% and 4%, respectively, of the global transgenic acreage and that the most frequently grown crops are GM soybean (61%), maize (23%), cotton (11%) and canola (5%).

Although few developing countries have released GM crop varieties, a preliminary analysis reveals that more than 20 developing countries are conducting research into the applications of GM crops. Although transgenic animals (especially mice) are used routinely in research (particularly in the medical field), no GM animals have yet been released on farms. A wide range of traits of potential interest to livestock producers have, however, been the subject of research; for example, the gene responsible for the production of growth hormone (which could be manipulated to increase growth rates), the phytase gene (which could reduce phosphorous emissions from pigs) and keratin genes (which could improve the wool of sheep).

The genetic modification of livestock has proceeded much more slowly than the genetic modification of crops for a variety of reasons, including the high costs, the inefficiency of the gene transfer techniques and the low reproductive rates of animals. Recombinant deoxyribonucleic acid (DNA) approaches have been used to promote the expression of desirable genes, to hinder the expression of undesirable genes, to alter specific genes and to inactivate genes so as to block specific pathways.

It is estimated that at least 30 enzymes produced by GM bacteria, yeasts and moulds are currently commercially available worldwide; many of these enzymes are used in the food industry. Genetic engineering has been used to introduce foreign genes into the animal genome or, alternatively, to knock out selected genes. Genes controlling growth were introduced into pigs to increase growth and improve carcass quality. Currently, research is underway to engineer resistance to diseases that affect the animals or that pose an indirect risk to human health, such as Marek's disease and salmonellosis in poultry, scrapie in sheep and mastitis in cattle.

Other studies have tried to increase the casein content of milk or to engineer animals that produce pharmaceutical or industrial chemicals in their milk or semen. No agricultural applications have yet proved commercially successful.

Nuclear transfer (NT) technology now provides an alternative route for cell-based transgenesis in domestic species, offering new opportunities for genetic modification. Livestock that produce human therapeutic proteins in their milk, that have organs suitable for xenotransplantation and that are resistant to diseases such as spongiform encephalopathies have been produced by NT from engineered cultured somatic cells.

CHARACTERISING GENETIC VARIABILITY

There is considerable genetic diversity in the livestock of developing countries, much of which controls traits that influence adaptability to harsh environments, productivity and susceptibility to disease and parasitism. However, little if any data on these genetic resources are available. Economic analysis can play an important role in ensuring that conservation efforts are appropriately focused.

The primary challenge facing conservationists is to identify sound reasons why society should preserve animals that livestock keepers have abandoned. Jabbar and Diedhiou show that the breeding practices and breed preferences of livestock keepers can successfully be determined by using research techniques such as the revealed preference hedonic approach. On the one hand, conservation cannot be achieved through a conventional breeding programme because the animals carrying the most advantageous traits cannot be easily identified; on the other hand, conservation cannot be achieved through biotechnology because the necessary technologies are either unavailable or uneconomic.

In livestock populations with a high degree of genetic variation, molecular markers are being increasingly used to study the distribution and patterns of genetic diversity. Global surveys indicate that 40% of domestic livestock breeds are at risk of extinction. Most of these breeds are found only in developing countries, and often little is known about them or their potential.

Rapid progress is being made in the preparation of dense microsatellite linkage maps to assist in the search for genetic traits of economic importance. These linkage maps can be used to develop strategies for marker assisted selection and marker assisted introgression that will meet the goals of breeding programmes in developing countries. Molecular markers have been widely used in the identification of genotypes and the 'genetic fingerprinting' of organisms. Genotype verification is used intensively to determine the parentage of domestic animals and to trace livestock products in the food chain back to the farm and animal of origin.

REPRODUCTIVE TECHNOLOGIES

The main objectives of using reproductive biotechnologies in livestock are to increase production, reproductive efficiency and rates of genetic improvement. Over the years, many options have become available for managing the reproduction of the major large and small ruminants. Artificial insemination (AI) and preservation of semen are the main technologies that are used extensively.

Assessing the fertilisation capacity of sperms, sexing sperms, synchronisation and fixed-time insemination, superovulation, embryo transfer (ET) and *in vitro* embryo production (IVEP) are additional techniques that can improve reproductive efficiency and pregnancy rates. Reproductive technologies can also be used to control reproductive diseases if procedures and protocols are accurately followed.

Artificial Insemination

The conception rate in field AI programmes in developing countries is very low, and therefore the desired effect in terms of animal improvement has not been achieved. Most semen banks still evaluate semen on the basis of sperm motility, even though significant advances have been made in techniques for semen evaluation. Although detailed guidelines are available regarding the processing, storage and thawing of cattle semen and buffalo semen, the processing and handling procedures in laboratories processing semen are often inadequate.

Only when farmers have access to considerably better technical and organisational facilities will AI become more effective. At present, the efficiency of the technology is limited by organisational and logistical constraints and by the failure to provide appropriate training for farmers. Several modifications of the technique have been suggested to increase the conception rate.

Synchronisation with different compounds, and the use of gonadotropin-releasing hormone (GnRH) followed seven days later by prostaglandin F2á (PGF2á) can synchronise oestrus and improves the conception rate. In this protocol giving injections of GnRH on day 0, PGF2á on day 7 and GnRH on day 9 is called the 'Ovsynch' programme and synchronises ovulation, permitting timed insemination. The ability to control ovarian follicular and corpus luteum development has allowed insemination in cattle to be timed without having to detect oestrus, and this has increased the net revenue per cow.

Embryo Transfer

One of the major reproductive technologies that can facilitate genetic improvement in cattle is ET. Unfortunately, commercial ET programmes are limited by the high variability in the ovarian follicular response to gonadotropin stimulation. Multiple ovulation and embryo transfer (MOET) takes AI one step further, in terms of both the possible genetic gains and the level of technical expertise and organisation required. In 2001, 450,000 embryos were transferred globally, mainly in dairy cattle, with 62% being transferred in North America and Europe, 16% in South America and 11% in Asia.

The main potential advantage of MOET for developing countries is that the elite females of local breeds can be identified, and bulls can be produced from them for use in a field programme of breed improvement. Zebu cattle and buffaloes in developing countries exhibit less consistent follicular dynamics after superovulation than *Bos taurus* in the developed world. However, over the last 10 to 15 years, the number of transferable embryos produced by zebu donors has increased from 2.4 to 5.8 embryos per flush in the late 1980s to 5.6 to 9.9 embryos per flush in 2000.

The use of ET has been less successful than envisaged for several reasons. The low reproductive efficiency, poor superovulatory responses, very low primordial follicle population and high incidence of atresia all contribute to low embryo production. In buffaloes, embryo recovery was initially less than one, but has subsequently improved to 2.6 with 1.4 transferable embryos per flush. After transferring buffalo embryos to recipients, the conception rate is only 16%. The poor success rates have limited the use of ET in buffaloes, which are the main dairy animals in developing countries in Asia, South-East Asia and the Mediterranean region.

In Vitro Production of Embryos

Since the birth of the first buffalo calf from an *in vitro* fertilised oocyte, a number of publications have described the effects of different protocols and media on oocyte and embryo development. Two extensive reviews have been published recently. However, the practical use of IVEP is limited by high production costs and the low overall efficiency under field conditions. High rates of maturation (70% to 90%), fertilisation (60% to 70%) and cleavage (40% to 50%), and moderate to low rates of blastocyst formation (15% to 30%) and calf production (10.5%) have been reported in the literature.

The efficiency of blastocyst production in buffaloes is much poorer than the 30% to 60% reported for cattle. Although viable buffalo blastocysts have been produced from ovaries obtained from abattoirs, the yield of transferable embryos remains low (15% to 39%). Embryos produced *in vitro* have led successfully to pregnancy and calf birth in buffalo, but the success rate is low. Therefore IVEP must be improved before it can be widely used in cattle and buffaloes in developing countries.

IMPROVING HEALTH THROUGH DEVELOPING VACCINES

Most biotechnologies related to health focus on the needs of the developed world, meaning that 90% of health research is devoted to the health problems of 10% of the world's population. Two main approaches are being used to develop vaccines using recombinant DNA technology. The first involves deleting genes that determine the virulence of the pathogen, thus producing attenuated organisms (non-pathogens) that can be used as live vaccines. Currently, this strategy is more effective against viral and bacterial diseases than against parasites. Attenuated live vaccines have been developed against the herpes viruses that cause pseudorabies in pigs and infectious bovine rhinotracheitis in cattle. A number of candidate *Salmonella* vaccines have also been produced. The second approach is to identify protein subunits of pathogens that can stimulate immunity.

The International Livestock Research Institute (ILRI) used this approach to develop a vaccine against *Theileria parva*, the parasite that causes East Coast fever in African cattle. A novel strategy for developing vaccines against bloodsucking parasites involves using components of the gut wall of the parasite that are not usually exposed to the immune system of the host. When the parasite feeds, it ingests antibodies induced by the vaccine, which destroy the gut wall and, consequently, kill the parasite. This strategy has been used successfully to develop a vaccine against the one-host tick *Boophilus microplus*.

Vaccination is one of the most effective and sustainable methods of controlling disease. Vaccines against parasitic diseases in Africa and viral diseases in Asia have been shown to control disease effectively and increase livestock productivity. A recent approach has been to use vaccines based on DNA. The use of DNA in vaccines is based on the discovery that injecting genes in the form of plasmid DNA can stimulate an immune response to the respective gene products. This immune response is a result of the genes being taken up and expressed by cells in the animal after injection. The live-vector and DNA vaccination systems could be manipulated further to enhance the immunity conferred by the gene products. Experimental studies have demonstrated that these vaccines can potentially induce appropriate and enduring immune responses.

This technology is, in principle, one of the simplest and yet most versatile methods of inducing both humeral and cellular immune responses, as well as protecting against a variety of infectious agents. However, although immune responses have been induced in a number of larger species, most of the information on the efficacy of DNA immunisation comes from studies of mice. An exhaustive review of the information available on the use of DNA vaccines in farm animals, including cattle, pigs and poultry, has identified the areas that need specific attention before this technology can be used routinely.

These areas include the delivery, safety and compatibility of plasmids in multivalent vaccines and the potential for using immune stimulants as part of a DNA vaccine. Korean scientists have developed a combined vaccine against pleuropneumonia, pneumonic pasteurellosis and enzootic pneumonia in swine. Molecular biology has been used to produce an improved vaccine against swine fever. In the Philippines, a vaccine has been developed that protects cattle and water buffalo against haemorrhagic septicaemia, which is the leading cause of death in these animals. The new vaccine provides improved protection at a very low cost.

DIAGNOSTICS AND EPIDEMIOLOGY

Advanced diagnostic tests that use biotechnology enable the agents causing disease to be identified and the impact of disease control programmes to be monitored more precisely than was previously possible. Molecular epidemiology characterises pathogens (viruses, bacteria, parasites and fungi) by nucleotide sequencing, enabling their origins to be traced. This is particularly important for epidemic diseases, in which pinpointing the source of the infection can significantly improve disease control. For example, the molecular analysis of rinderpest viruses has been vital in determining the lineages circulating in the world and instrumental in aiding the Global Rinderpest Eradication Programme. Enzyme-linked immunosorbent assays have become the standard means of diagnosing and monitoring many animal and fish diseases worldwide, and the PCR technique is especially useful in diagnosing livestock disease.

Many diagnostic techniques currently used in developing countries are cumbersome and unsuitable for low-resource settings. Molecular diagnostic technologies that are either already in use or being tested in low-income regions include polymerase chain reaction (PCR), monoclonal antibodies and recombinant antigens. These technologies can be modified to facilitate their application in the developing world. Simple hand-held devices that rely on the binding specificity of monoclonal antibodies or recombinant antigens to diagnose infection may be easily adapted for use in settings without running water, refrigeration or electricity. Molecular characterisation of the virus serotypes causing foot and mouth disease has helped in the vaccination and control programmes in Asia.

In Japan and Taiwan, DNA testing is being used to diagnose hereditary weaknesses of livestock. One test looks for the presence of the gene responsible for porcine stress syndrome in pigs. Pigs with this gene tend to produce pale poor-quality meat because of their reaction to the stress of transport and slaughter. Pigs with this gene can now be excluded from breeding programmes, so the gene will become less common. In addition, DNA testing is being used in Japan to check for the gene that causes leucocyte adhesion deficiency in Holstein cattle. Cattle with this condition suffer from gum disease, tooth loss and stunted growth. They usually die before they are one year old.

By using DNA testing, carriers can be identified and eliminated from breeding herds. Bulls used for breeding can also be tested to make sure that they are not carriers. Another DNA test identifies a gene that leads to anaemia and retarded growth in Japanese Black cattle.

NUTRITION AND FEED UTILISATION

The shortage of feed in most developing countries and the increasing cost of feed ingredients mean that there is a need to improve feed utilisation. Aids to animal nutrition, such as enzymes, probiotics, single-cell proteins and antibiotics in feed, are already widely used in intensive production systems worldwide to improve the nutrient availability of feeds and the productivity of livestock. Gene-based technologies are being increasingly used to improve animal nutrition, either through modifying the feeds to make them more digestible or through modifying the digestive and metabolic systems of the animals to enable them to make better use of the available feeds. Feeds derived from GM plants (a quarter of which are now grown in developing countries), such as grain, silage and hay, have contributed to increases in growth rates and milk yield. Genetically modified crops with improved amino acid profiles can be used to decrease nitrogen excretion in pigs and poultry. Increasing the levels of amino acids in grain means that the essential amino acid requirements of pigs and poultry can be met by diets that are lower in protein. Metabolic modifiers have also been used to increase production efficiency (weight gain or milk yield per feed unit), improve carcass composition (meat-fat ratio), increase milk yield and decrease animal fat.

The use of recombinant bovine somatotropin (rBST) in dairy cows increases both milk yield and production efficiency and decreases animal fat. In the USA, the use of rBST typically increases milk yield by 10% to 15%. Although trials conducted in developing countries have reported a similar percentage increase, this increase is not significant because of the low milk yields and the high cost-benefit ratio. However, rBST is being used commercially in 19 countries where the economic returns make its use worthwhile. A porcine somatotropin has been developed that increases muscle growth and reduces body-fat deposition, resulting in pigs that are leaner and of greater market value.

CONSTRAINTS ON APPLYING THE TECHNOLOGY

The application of new molecular biotechnologies and new breeding strategies to the livestock breeds used in smallholder production systems in developing countries is constrained by a number of factors. In the developing world, poverty, malnutrition, disease, poor hygiene and unemployment are widespread, and biotechnologies must be able to be applied in this context.

Over the last few decades, the green revolution has brought comparative prosperity to farmers with land, but the majority of farmers, who are landless or marginal farmers and subsist only on livestock, have been neglected and remain poor.

The major constraints on applying biotechnologies have been enumerated by Madan and include:

- The absence of an accurate and complete database on livestock and animal owners so that programmes can be implemented
- The biodiversity present within species and breeds in agro-ecological systems
- The fact that models of biotechnological intervention differ distinctly between developed and developing economies
- The fact that many animal species and breeds are unique to the developing world; each has its own distinct developmental, production, disease resistance and nutrient utilisation characteristics
- The lack of trained scientists, technicians and fieldworkers to develop

and apply the technologies, both in the government and in the private sectors

- The absence of an interface between industry, universities and institutions, which is necessary to translate technologies into products
- The inability to access technologies from the developed world at an affordable price in order to make a rightful, positive and sustainable contribution to livestock production and the economic welfare of farmers
- The high cost of technological inputs such as materials, biologicals and equipment
- The failure to address issues of biosafety and to conduct risk analyses of new biologicals, gene products, transgenics and modified food items, and, above all
- The negligible investment in animal biotechnology.

The critical issues affecting livestock productivity have recently been re-examined. Research that aims to enhance productivity and sustainability should focus on improving livestock feeds and nutrition, improving animal health, managing natural resources relating to the livestock sector, assesing the impact of technological interventions, and strengthening the capacity of the national agricultural research systems of developing countries.

Furthermore, the potential production capacity and contribution of livestock to the economy are still not being achieved in developing countries because the transfer, adaptation and adoption of technology is hampered by the lack of a clear policy for livestock development that is conducive to the introduction of new proven technology and by the lack of information flow from and to decision makers. In developing countries, there is a wealth gap between urban and rural areas, which persists and may even be widening; the rural-urban divide also tends to be reflected in education and health indicators. In addition, women in rural (and urban) areas who are predominantly involved in animal husbandry have higher illiteracy rates than men. A survey of 21 African countries recently highlighted the substantial disparities in primary schooling between urban and rural areas, in favour of urban dwellers. Special attention must be given to the knowledge and information needed to enable rural people to apply biotechnology. There is a need to identify alternative delivery systems for animal healthcare and to propose new roles for the state and the private sector in service delivery.

BUILDING CAPACITY

Owing to the constraints outlined above, the economic benefits of animal biotechnology cannot be realised without a conscious, sustained, holistic, multi-stakeholder, participatory approach. There is a great need to ensure that capacity is not just created but also is retained and enhanced. Capacity-building activities must be carried out at all levels: the awareness of policy and decision makers must be raised, the necessary legal and regulatory frameworks must be initiated, the technical and regulatory capacities must be enhanced and institutions may need to be overhauled. More importantly, it is necessary to assess and deploy competent operators and institutional capacity continuously so that, as biotechnology advances, the procedures required for its safe use can be constantly evaluated, upgraded and applied. This is a daunting task, but it can be achieved through firm commitment and partnerships.

FUNDING TO IMPLEMENT TECHNOLOGY

Developing and commercialising improved technologies in most developing countries has been the responsibility of the public sector, and technology has been disseminated freely. This situation will have to continue if superior genetics, diagnostics and vaccines are to be delivered. However, research and almost all commercial development of biotechnology in the developed world are being driven largely by the private sector. The global trends in funding for research and development and production do not address the concerns, needs and opportunities of the developing world. Developing countries are finding it increasingly expensive to access and use new technologies. There is limited private-and public-sector investment in animal health and production, particularly in relation to modern biotechnologies that are 'resource hungry'. Although several discoveries have been made in laboratories in the developing world, in most cases these have not been converted into useful technologies or products. The key potential users–resource-poor often illiterate farmers with a limited knowledge base–do not feel that applying these technologies is worth the effort, cost and risk involved. This is mainly because there is no agency or industry that can scale up and package the technology. Also, in the developed world, there is an economic incentive to market biotechnological services and products; this is lacking in the developing world because of the limited purchasing power of resource-poor stakeholders.

Research in biotechnology in recent years has also been motivated by economic considerations, and little research is conducted in the developing world because of the probable lack of returns on the investment. For understandable reasons, current funding policies in developing countries focus on areas that will yield practical benefit in the short term. In determining future policy, policy-makers and funding bodies must not lose sight of the substantial benefits that can be gained in the longer term by investing in strategic research into vaccine development. Adequate multi-institutional (national and international) support through an international donor consortium is needed to develop cost-effective, cheap and easily adaptable biotechnological products.

nt spent by international agencies on animal biotechnology in developing countries is currently very low and constitutes only a small percentage of the total spending on agriculture. The World Bank, the Food and Agriculture

Organization, the Consultative Group on International Agricultural Research, the United Nations Development Programme, the United States Agency for International Development, the Swedish International Development Cooperation Agency, the International Development Research Centre, the Asian Development Bank and other donor and funding agencies have to designate a higher percentage of funds to the livestock sector. It has been convincingly shown that investing in livestock has a dramatic and far-reaching impact on the human development index. This is a strong argument in favour of investing heavily in animal production and health biotechnologies in order to bring economic prosperity, nutritional security, rural development and health improvements to poor populations in the developing world.

Although animal production is being changed significantly by advances made in thousands of biotechnology laboratories around the world, benefits are reaching the developing world in only a few areas of conservation, animal improvement, healthcare (including diagnosis and control of disease) and the augmentation of feed resources. Adopting biotechnology has resulted in distinct benefits in terms of animal improvement and economic returns to the farmers. Over the past decade, the ILRI has focused on biotechnological applications, especially in Africa, and several developing countries now have multi-institutional programmes to develop and apply biotechnology. The developing world will have to respond to the many genebased technologies now being developed with a sense of commitment, trained manpower, infrastructure and funding.

5

Enzymes

The use of enzymes in the diagnosis of disease is one of the important benefits derived from the intensive research in biochemistry since the 1940's. Enzymes have provided the basis for the field of clinical chemistry. It is, however, only within the recent past few decades that interest in diagnostic enzymology has multiplied. Many methods currently on record in the literature are not in wide use, and there are still large areas of medical research in which the diagnostic potential of enzyme reactions has not been explored at all. Enzymes are biological catalysts responsible for supporting almost all of the chemical reactions that maintain animal homeostasis. Because of their role in maintaining life processes, the assay and pharmacological regulation of enzymes have become key elements in clinical diagnosis and therapeutics.

The macromolecular components of almost all enzymes are composed of protein, except for a class of RNA modifying catalysts known as ribozymes. Ribozymes are molecules of ribonucleic acid that catalyse reactions on the phosphodiester bond of other RNAs. Enzymes are found in all tissues and fluids of the body. Intracellular enzymes catalyse the reactions of metabolic pathways. Plasma membrane enzymes regulate catalysis within cells in response to extracellular signals, and enzymes of the circulatory system are responsible for regulating the clotting of blood. Almost every significant life process is dependent on enzyme activity.

PROPERTIES OF ENZYMES

Following are the two most important properties of enzymes:

1. Remarkable specificity provided due to enzyme's active site for substrate binding and subsequent chemical transformation,
2. Often a phenomenal acceleration in the rate of reaction, which can reach a level of 10^{10} to $10^{2}3$ times the non-enzymatic rates. In absolute terms, the catalytic events due to an enzyme molecule can be as slow as one event per minute to 10^5 events per second.

A typical enzyme, having a molecular mass of 50,000 (M_r 50K) is comprised of 450 amino acid residues, of which ~ 400 residues involve 19 chiral-L-amino acids and the remaining ~50 residues consist of glycine (a non-chiral or symmetric amino acid; chiral means asymmetric).

Out of 450 amino acids of an enzyme, it is the 400 chiral amino acid residues, which provide an asymmetric micro-environment for the formation of an active site for substrate binding and subsequent chemical transformation.

Parameters Affecting Enzymatic Reactions

The rate of an enzymatic reaction depends on several factors including the reaction itself, the nature of enzyme involved, and more importantly the concentrations of substrate (S), product (P) and enzyme (E). For instance, if the concentration of the substrate is slowly increased, keeping other things constant, the rate of reaction increases till it reaches a maximum, where further increase in substrate concentration does not lead to any further increase in the rate of reaction.

This maximum rate of reaction under a set of constant conditions is described as V_x. Besides substrate concentration, several other parameters have been suggested to describe fully an enzymatic reaction.

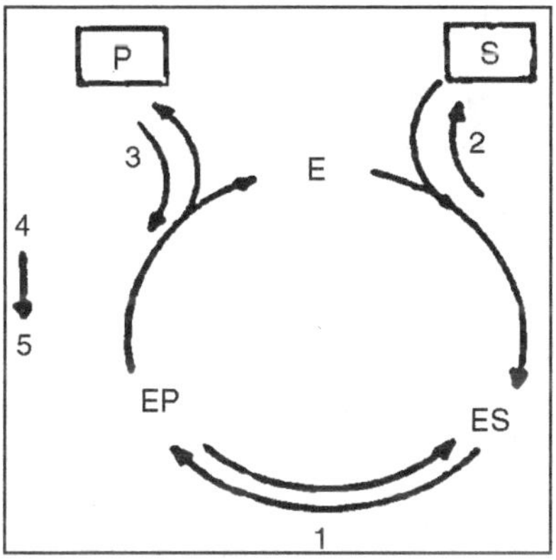

Fig. Different Parameters, Enzymatic Reactions

- K_{cat}: the higher the better
- $K_{cat}/K_M.K_M$ (K_s) Low K_M causes substrate inhibition
- low K_p causes products inhibition
- 1 Umg^{-1} ~ ($K_{cat}{}^2s^{-1}$)
- 1 mmol product per day

These parameters include the following:

- Dissociation constants of substrate S and product P (K_s and K_p);
- Specific activity or the turnover number, which is quantified by k_{cat} and is obtained, when V_x is divided by. enzyme concentration [E];

this represents the number of substrate molecules processed per enzyme molecule per second and its value is generally 1000, although it can be much greater;

- Michaelis constant (K_m), which represents the substrate concentration, at which the rate of reaction is half its maximum rate V_x described above;
- Specificity or catalytic efficiency of enzyme, determined by maximum specificity constant (k_{cat}/K_m).

Enzyme Kinetics Showing Relationship Between the Rate of Enzyme Reaction and the Substrate Concentration, Km is the Substrate Concentration at which the Reaction Rate is Half Maximal

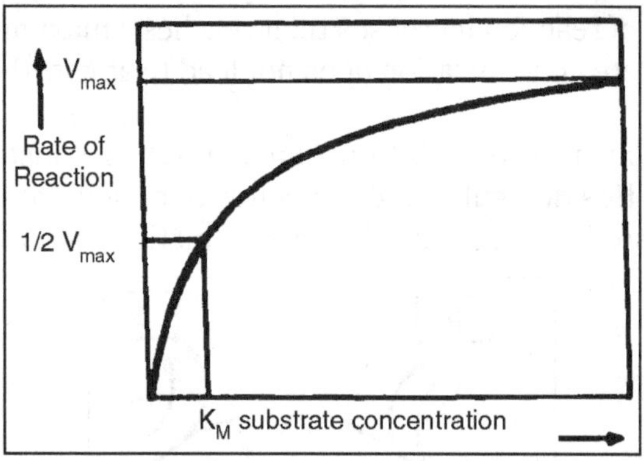

Fig. Enzymatic Reactions

Among the above parameters, the turnover number {k_{cat} (s^{-1})}, which describes the decomposition of enzyme-substrate complex (E-S) is a first order rate constant, and the catalytic efficiency {k_{cat}/K_m ($M^{-1}s^{-1}$)}' which describes the rate of product formation is a second order rate constant. The values of catalytic efficiencies of enzymes lie in the range of 10^5 M^{-1} s^{-1} (*e.g.* aspartate aminotransferase) to 10^9 M^{-1} s^{-1} (*e.g.* fumarase) and cluster around a mean value of $\sim 10^7$ s^{-1}, which is common for many hydroxylases.

The relationship of these two rate constants and that of a rate constant for the uncatalysed reaction (k_{non}) with the rate of product formation is shown in the following three sets of equations:

1. $ES \rightarrow E + S$

   $d[P]/dt = k_{cat}$ [ES]
2. $E + S \rightarrow E + P$

   $d[P]/dt = k_{cat}/k_m$ [E] [S]
3. $E \rightarrow P$

   $d[P]/dt = k_{non}$ [S]

By comparing the rate constant of an uncatalysed reaction (k_{non}) with that of catalysed reaction (k_{cat}) one can calculate the increase in reaction rate that an enzyme produces; this is described as rate enhancement and is calculated as k_{cat}/k_{non}. Another parametre termed as catalytic proficiency is, calculated as the ratio $(k_{cat}/K_m)/k_{non}$.

Ideally, for the economic' production at the industrial scale, the values of dissociation constants, K_s and K_p should be low, *i.e.*, the enzyme should be subject to minimum substrate and product inhibition. In constrast, the values of turnover number (k_{cat}) and catalytic efficiency (k_{cat}/K_m) should be high to allow high rate of product formation. Specificity of an enzyme for a substrate is also important, because for general-purpose catalysis, enzymes with broader specificity would prove useful, although enzymes with narrow specificity may be more efficient.

Medium for Enzymes Aqueous vs Organic Solvent

Most enzymes function in their natural aqueous media, which limit their use in industry for the production of speciality chemicals and polymers, for which organic solvents are the more suitable reaction media, provided the enzyme can function in such a medium. Normally, the enzymes get denatured in aqueous-organic mixture, leading many to believe that they will do so in pure organic solvents also.

However, it was shown during 1985-2000, that various crystalline enzymes retain their structure even in anhydrous solvents. In fact, some enzymatic reactions work better in organic solvents. Sometimes, enzymes may be more stable and specific to the organic solvents, thus providing opportunities for their use in industries, where aqueous solvents become a limitation.

There are several factors, which facilitated the use of organic solvents in enzymatic reactions:

- Transesterification (due to addition of alcohol), aminolysis (due to addition of amines) and thiotransesterification (due to addition of thiols) are suppressed in aqueous solution and are important transformations that are needed in industry.
- Hydrophobic solvents are superior to hydrophilic solvents, because the latter will strip from the enzyme molecules, the tightly bound water (which is essential for catalytic activity).
- No specific pH is needed, while using organic solvents.
- The enzymes in organic solvents can be lyophilized to give enzyme powder without any damage to the enzyme activity. In brief we can say that strategies are being developed to have more efficient enzymes in organic solvents, their activities sometimes being comparable to that in aqueous medium.

Selectivity of Enzymes

One of the important properties of enzymes is to select from a mixture (racemic pairs, prochiral or meso-compounds or diastereometric mixtures), a compound, which is enantiomerically pure (enantiomers are a class of isomers that are mirror images of each other and occur in mixtures).

This is described as enzymatic resolution of substrates and can be achieved by anyone of the following three methods:

1. The enzyme esterifies only a single enantiomer of a racemic substrate and thus allows its separation. Enol esters can be used as transesterification reagents to force the enzymatic process to proceed irreversibly in the desired forward direction.
2. Enzymatic resolution has also been realised by certain hydrolytic enzymes through the use of additional racemization catalysts, which keep on converting the enantiomer that is not a substrate for the enzyme into the enantiomer, which is the substrate. In situ racemization thus gradually increases the overall concentration of the substrate recognized by the enzyme, leading to higher product yield.
3. Enzymatic resolution may involve efficient protecting-group strategies. For instance, some lipases have been shown to catalyse reactions that provide temporary protection of amines. Some examples of 'biocatalytic processes, which involve stereoselectivity of enzymes for enzymatic resolution.

Types of Enzymes

The enzymes that have been exploited for organic synthesis, and the reactions catalysed by each of these types are summarized in. Enzymes have been classified into the following six classes, and a standard nomenclature has been suggested for them by the 'Commission on Biochemical Nomenclature' of the 'International Union of Biochemistry. According to standard nomenclature, each enzyme is assigned' a four number code following the abbreviation EC (for Enzyme Commission). For instance, EC1.1.1.3 means an enzyme number 3 belonging to class 1 (oxidoreductase), subclass 1 and sub-subclass 1.

Oxidoreductases (Class 1)

Oxidoreductases facilitate redox reactions involving transfer of hydrogen or oxygen atoms between molecules.

They include oxidases (e^- transfer to oxygen), oxygenases (oxygen atom transfer from molecular oxygen or O_2) prexidases (e^- transfer to peroxides) and dehydrogenases (hydride transfer). Glucose oxidase (EC1.1.3.4) is a popular example of this class of enzymes.

Transferases (Class 2)

Transferases catalyse the transfer of an atom (other than oxygen or hydrogen) or a group of atoms like acyl-, alkyl-, glycosyl- etc. For instance, glycosyl-transferases catalyse transfer of glycosyl group, sulphotransferases catalyse transfer of sulphate group and aminotransferases catalyse transfer of amino group. Aspartate aminotransferase (EC2.6.1.1) is a suitable example of this class of enzymes.

Hydrolases (Class 3)

Hydrolases catalyse hydrolytic reactions or their reverse reactions (condensation) and include esterases, amidases, lipases, proteases, glycosidases, etc. Amidases also help in peptide synthesis (by condensation of acid and amino groups) and esterases also help in the formation of esters (by' condensation between acid and alcohol). Similarly, glycohydrolases are used for the synthesis of glycoproteins. Most of these hydrolytic enzymes are important for' the pharmaceutical industry. Chymosin or rennin (EC3.2.3.4) is a suitable example of this class of enzymes.

Lyases (Class 4)

Lyases catalyse reactions involved in the removal of a group of atoms from the substrate molecule. These enzymes include aldolases, decarboxylases, dehydratases and some pectinases. Histidine ammonia lyase (EC 3.2.1.3) is a suitable example of this class.

Isomerases (Class 5)

Isomerases catalyse the formation f isomers of substrate molecules and therefore, include epimerases, racemases, and intramolecular transferases. Xylose isomerase (EC5.3.1.5) is a suitable example of this class.

Ligases or Synthetases (Class 6)

Ligases catalyse the formation of covalent bonds between two substrate molecules with the help of ATP or GP molecules, which provide the energy for the reaction. Glutathione synthase (EC6.3.2.3) is a suitable example of this class.

Biomimetic Catalysts

Biomimetic catalysts are the synthetic organic or inorganic compounds which perform the same function as the natural enzymes. These biomimetic catalysts may operate under harsher solvent and temperature conditions and may be more robust in terms of lifetime, even if they may be unsuitable for high throughput. Some of the processes that can be catalyzed by biomimetic catalysts include macrocyclization of natural products, production of analogues of hydrogenase or nitrogenase catalysts or the photosynthetic splitting of water.

However, sometimes natural enzymes should be preferred over synthetic reagents and catalysts, as in case of the following:

- Lipases and other hydrolases, have an advantage in kinetic resolutions of intermediates,
- Peniallin acylases are useful in β-lactam antibiotic industry
- Aldolases and transketolases help in the synthesis of oligosaccharides due to aldol condensation.

Table. Enzyme Commonly Used in Organic Synthesis

Enzymes	Reactions
1. Esterases, lipases	Ester hydrolysis, ester formation
2. Amidases (proteases, acylases)	Amide hydrolysis, amide formation
3. Dehydrogenases	Oxidoreduction of alcohols and ketones
4. Oxidases	Oxidation
5. Peroxidases	Oxidation, epoxidation, halohydration
6. Kinases	Phosphorylation (ATP dependent)
7. Aldolases, transketolases	Aldol reaction (C-C bond)
8. Glycosidases, glycosyltransferases	Glycosidic bond formation
9. Phosphorylases, phosphatases	Formation and hydrolysis of phosphate
10. Sulphotransferases	Formation of sulphate esters
11. Transaminases	Amino acid synthesis (C-N band)
12. Hydrolases	Hydrolysis
13. Isomerases, lyases, hydratases	Isomerization, addition, elimination, or replacement of a group

Extremozymes

The use of biocatalysis in chemical reactions is generally restricted to mild conditions of pH, temperature, pressure and aquatic medium. However, enzymes isolated from different living organisms inhabiting ecosystems with extreme conditions of pH, temperature, pressure and solvent, are not constrained with these mild conditions needed for majority of enzymes.

These enzymes, therefore, provide information to improve stability, activity and specificity of many enzymes used for industrial biocatalysis. It is now known that enzymes can function at temperatures as high as 140°C and as low as below freezing point.

Organisms living in these conditions are described as extremophiles and the enzymes that function under these conditions are therefore described as extremozymes. Most of these extremophiles belong to archaea group of bacteria.

Extremozymes are broadly classified as follows:

- Extremely psychophilic enzymes, which function at temperatures approaching the freezing point of water; for instance, subtilisin S41 from the psychophile Bacillus S41 differs from other subtilisins in several respects, suggesting structural differences between these psychophilic enzymes and those extracted from mesophiles.

- Extremely halophilic enzymes function better in salt solutions. For instance, aspartate aminotransferase from Haloferas mediterranes is inactivated rapidly at room temperature in low salt solutions, but does not denature even at 78.5°C in 3.3M KCI suggesting that the enzyme is more stable in higher salt concentration. A protease from Halobacterium halobium is also similarly affected.
- Extremely thermophilic enzymes function at high temperatures and several examples of these enzymes are available now. (DNA polymerase from Thermus aquaticus and Pyrococcus furiosus used for PCR provide one such example).
- Extremely barophilic enzymes have been isolated from microbes living in deep sea which are both barophils and psychophiles.

A number of genes encoding extremozymes (*e.g.* ferredoxin, DNA polymerase, amylase, dehydrogenase, etc.) have been isolated from thermophilic archaea (*P. furiosus, P. woesii* and *Thermococcus littoralis*), cloned and expressed in *E. coli*.

Three dimensional structures of some extremozymes have provided opportunities to provide or design more stable versions of enzymes, that can function in extreme salinity, pH, temperatures and non-aqueous media. Selectivity of extremozymes (enantioselectivity, regioselectivity and chemoselectivity) can also be adjusted through judicious use of particular enzymes in organic solvents or water/solvent mixtures.

Cofactor-Dependent Enzymes and Cofactor Regeneration

A number of enzymatic reactions require cofactors, some of the common inorganic cofactors being iron/sulphur clusters and metalloporphyrins. These cofactors are costly and therefore can not be used in industry, unless regenerated as a reaction by-product.

Some of the cofactors which can be recycled effectively include the following:

- Hucleoside triphosphates, such as ATP in phosphoryl-transfer reactions,
- Nicotinamide adenine dinucleotide (NAD) and its 3' phosphate (NADP) in oxidoreductions,
- Acetyl CoA in acyl transfer reactions,
- 3'-phosphoadenosine- 5'- phosphosulphate (PAPS) in the formation l of sulphate esters
- Sugar nucleotides in glycosyl transfer reactions. Many cofactor dependent reactions involving regeneration of cofactor have been used

for industry. However regeneration of cofactor may require additional enzymes, whose number may be reduced by using bifunctional enzymes.

Modular Enzymes

A modular enzyme is a multicomponent system or an assembly line having several component enzyme domains (also called modules). The order of enzymatic domains or modules in these multicomponent systems specifies which monomer substrates are activated, condensed and elongated. Therefore, the individual domains or components can be interchanged or replaced with functionally distinct analogues from related systems, inorder to obtain products with altered desired structure.

The polyketide synthase (PKS) and non-ribosomal peptide synthetase (NRPS) are two such multicomponent systems, which can be modified by gene shuffling to provide designed diversity in complex natural products (*e.g.* antibiotics like penicillins, vancomycin and erythromycin).

Following three general classes of modular biocatalysts can be identified:

1. Enzymes in which catalysis and substrate specifities are separable,
2. Multi substrate enzymes, in which binding sites for individual substrates are modular,
3. Enzymes, where an intermediate is transferred from one active site to another in a metabolic sequence without equilibrating with the bulk fluid.

Coenzymes

The chemical reactivity and specificity of an enzyme largely depends on its catalytic site, made up of the side chains of chiral amino acids (glycine is the only non-chiral amino acid, the remaining 19 amino acids are chiral in nature). However, sometimes the side chains of chiral amino acids are insufficient and the enzyme may need a non-polypeptide molecule to bind on its surface thus enabling it to function as an enzyme. These non-polypeptide molecules serve as ligands and are called coenzymes. The coenzymes are therefore defined as small molecules that are associated with enzymes, and are essential for the activity of enzymes. Some of these coenzymes are involved in group transfers.

Table. Some Coenzymes Involved in Group Transfer Reactions

Coenzymes	Group transferred
ATP	Phosphate
NADH, NADPH	Hydrogen and electron
Coenzyme A	Hydride ion = H+, acetyl
Biotin	Carboxyl
S – adenosylmethionine	Methyl

ROLE OF COENZYMES

The functional role of coenzymes is to act as transporters of chemical groups from one reactant to another. The chemical groups carried can be as simple as the hydride ion ($H^+ + 2e^-$) carried by NAD or the mole of hydrogen carried by FAD; or they can be even more complex than the amine ($-NH_2$) carried by pyridoxal phosphate.

Since coenzymes are chemically changed as a consequence of enzyme action, it is often useful to consider coenzymes to be a special class of substrates, or second substrates, which are common to many different holoenzymes. In all cases, the coenzymes donate the carried chemical grouping to an acceptor molecule and are thus regenerated to their original form. This regeneration of coenzyme and holoenzyme fulfills the definition of an enzyme as a chemical catalyst, since (unlike the usual substrates, which are used up during the course of a reaction) coenzymes are generally regenerated.

ENZYME RELATIVE TO SUBSTRATE TYPE

Although enzymes are highly specific for the kind of reaction they catalyze, the same is not always true of substrates they attack. For example, while succinic dehydrogenase (SDH) always catalyzes an oxidation-reduction reaction and its substrate is invariably succinic acid, alcohol dehydrogenase (ADH) always catalyzes oxidation-reduction reactions but attacks a number of different alcohols, ranging from methanol to butanol. Generally, enzymes having broad substrate specificity are most active against one particular substrate. In the case of ADH, ethanol is the preferred substrate.

Enzymes also are generally specific for a particular steric configuration (optical isomer) of a substrate. Enzymes that attack D sugars will not attack the corresponding L isomer. Enzymes that act on L amino acids will not employ the corresponding D optical isomer as a substrate. The enzymes known as racemases provide a striking exception to these generalities; in fact, the role of racemases is to convert D isomers to L isomers and vice versa. Thus racemases attack both D and L forms of their substrate.

As enzymes have a more or less broad range of substrate specificity, it follows that a given substrate may be acted on by a number of different enzymes, each of which uses the same substrate(s) and produces the same product(s). The individual members of a set of enzymes sharing such characteristics are known as isozymes. These are the products of genes that vary only slightly; often, various isozymes of a group are expressed in different tissues of the body.

The best studied set of isozymes is the lactate dehydrogenase (LDH) system. LDH is a tetrameric enzyme composed of all possible arrangements of two different protein subunits; the subunits are known as H (for heart) and M (for skeletal muscle). These subunits combine in various combinations leading to 5 distinct isozymes. The all H isozyme is characteristic of that from heart tissue, and the all M isozyme is typically found in skeletal muscle and liver. These isozymes all catalyze the same chemical reaction, but they exhibit differing degrees of efficiency. The detection of specific LDH isozymes in the blood is highly diagnostic of tissue damage such as occurs during cardiac infarct.

ENZYME-SUBSTRATE INTERACTIONS

The favoured model of enzyme substrate interaction is known as the induced fit model. This model proposes that the initial interaction between enzyme and substrate is relatively weak, but that these weak interactions rapidly induce conformational changes in the enzyme that strengthen binding and bring catalytic sites close to substrate bonds to be altered. After binding takes place, one or more mechanisms of catalysis generates transition- state complexes and reaction products.

The possible mechanisms of catalysis are four in number:

1. *Catalysis by Bond Strain*: In this form of catalysis, the induced structural rearrangements that take place with the binding of substrate and enzyme ultimately produce strained substrate bonds, which more easily attain the transition state. The new conformation often forces substrate atoms and bulky catalytic groups, such as aspartate and glutamate, into conformations that strain existing substrate bonds.
2. *Catalysis by Proximity and Orientation*: Enzyme-substrate interactions orient reactive groups and bring them into proximity with one another. In addition to inducing strain, groups such as aspartate are frequently chemically reactive as well, and their proximity and orientation towards the substrate thus favours their participation in catalysis.
3. *Catalysis Involving Proton Donors (Acids) and Acceptors (Bases)*: Other mechanisms also contribute significantly to the completion of catalytic events initiated by a strain mechanism, for example, the use of glutamate as a general acid catalyst (proton donor).
4. *Covalent Catalysis*: In catalysis that takes place by covalent mechanisms, the substrate is oriented to active sites on the enzymes in such a way that a covalent intermediate forms between the enzyme or coenzyme and the substrate. One of the best-known examples of this mechanism is that involving proteolysis by serine proteases, which include both digestive enzymes (trypsin, chymotrypsin, and elastase) and several enzymes of the blood clotting cascade. These proteases contain an active site serine whose R group hydroxyl forms a covalent bond with a carbonyl carbon of a peptide bond, thereby causing hydrolysis of the peptide bond.

Chemical Reactions and Rates

According to the conventions of biochemistry, the rate of a chemical reaction by the number of molecules of reactant(s) that are converted into product(s) in a specified time period. Reaction rate is always dependent on the concentration of the chemicals involved in the process and on rate constants that are characteristic of the reaction.

For example, the reaction in which A is converted to B is written as follows:

$$A \rightarrow B$$

The rate of this reaction is expressed algebraically as either a decrease in the concentration of reactant A:

$$-[A] = k[B]$$

or an increase in the concentration of product B:

$$[B] = k[A]$$

In the second equation (of the 3 above) the negative sign signifies a decrease in concentration of A as the reaction progresses, brackets define concentration in molarity and the k is known as a rate constant. Rate constants are simply proportionality constants that provide a quantitative connection between chemical concentrations and reaction rates.

Each chemical reaction has characteristic values for its rate constants; these in turn directly relate to the equilibrium constant for that reaction. Thus, reaction can be rewritten as an equilibrium expression in order to show the relationship between reaction rates, rate constants and the equilibrium constant for this simple case. The rate constant for the forward reaction is defined as k_{+1} and the reverse as k_{-1}.

At equilibrium the rate (v) of the forward reaction (A → B) is, by definition, equal to that of the reverse or back reaction (B → A), a relationship which is algebraically symbolized as:

$$v_{forward} = v_{reverse}$$

where, for the forward reaction:

$$v_{forward} = k_{+1}[A]$$

and for the reverse reaction:

$$v_{reverse} = k_{-1}[B]$$

In the above equations, k_{+1} and k_{-1} represent rate constants for the forward and reverse reactions, respectively. The negative subscript refers only to a reverse reaction, not to an actual negative value for the constant. To put the relationships of the two equations into words, we state that the rate of the forward reaction [$v_{forward}$] is equal to the product of the forward rate constant k_{+1} and the molar concentration of A. The rate of the reverse reaction is equal to the product of the reverse rate constant k_{-1} and the molar concentration of B. At equilibrium, the rate of the forward reaction is equal to the rate of the reverse reaction leading to the equilibrium constant of the reaction and is expressed by:

$$K_{eq} = \frac{K_{+1}}{K_{-1}} = \frac{[B]}{[A]}$$

This equation demonstrates that the equilibrium constant for a chemical reaction is not only equal to the equilibrium ratio of product and reactant concentrations, but is also equal to the ratio of the characteristic rate constants of the reaction.

Chemical Reaction Order

Reaction order refers to the number of molecules involved in forming a reaction complex that is competent to proceed to product(s). Empirically, order is easily determined by summing the exponents of each concentration term in the rate equation for a reaction. A reaction characterized by the conversion of one molecule of A to one molecule of B with no influence from any other reactant or solvent is a first-order reaction.

The exponent on the substrate concentration in the rate equation for this type of reaction is 1. A reaction with two substrates forming two products would a second-order reaction. However, the reactants in second– and higher– order reactions need not be different chemical species. An example of a second order reaction is the formation of ATP through the condensation of ADP with orthophosphate:

$$ADP + H_2PO_4^- \leftrightarrow ATP + H_2O$$

For this reaction the forward reaction rate would be written as:

$$v_{forward} = k_1[ADP][H_2PO_4^-]$$

ALLOSTERIC ENZYMES

In addition to simple enzymes that interact only with substrates and inhibitors, there is a class of enzymes that bind small, physiologically important molecules and modulate activity. These are known as allosteric enzymes; the small regulatory molecules to which they bind are known as effectors. Allosteric effectors bring about catalytic modification by binding to the enzyme at distinct allosteric sites, well removed from the catalytic site, and causing conformational changes that are transmitted through the bulk of the protein to the catalytically active site(s).

The hallmark of effectors is that when they bind to enzymes, they alter the catalytic properties of an enzyme's active site. Those that increase catalytic activity are known as positive effectors. Effectors that reduce or inhibit catalytic activity are negative effectors.

Most allosteric enzymes are oligomeric (consisting of multiple subunits); generally they are located at or near branch points in metabolic pathways, where they are influential in directing substrates along one or another of the available metabolic paths. The effectors that modulate the activity of these allosteric enzymes are of two types.

Those activating and inhibiting effectors that bind at allosteric sites are called heterotropic effectors. (Thus there exist both positive and negative heterotropic effectors.) These effectors can assume a vast diversity of chemical forms, ranging from simple inorganic molecules to complex nucleotides such as cyclic adenosine monophosphate (cAMP). Their single defining feature is that they are not identical to the substrate. In many cases the substrate itself induces distant allosteric effects when it binds to the catalytic site.

Substrates acting as effectors are said to be homotropic effectors. When the substrate is the effector, it can act as such, either by binding to the substrate-binding site, or to an allosteric effector site. When the substrate binds to the catalytic site it transmits an activity-modulating effect to other subunits of the molecule. Often used as the model of a homotropic effector is hemoglobin, although it is not a branch-point enzyme and thus does not fit the definition on all counts.

There are two ways that enzymatic activity can be altered by effectors: the V_{max} can be increased or decreased, or the K_m can be raised or lowered. Enzymes whose K_m is altered by effectors are said to be K-type enzymes and the effector a K-type effector. If V_{max} is altered, the enzyme and effector are said to be V-type. Many allosteric enzymes respond to multiple effectors with V-type and K-type behaviour. Here again, hemoglobin is often used as a model to study allosteric interactions, although it is not strictly an enzyme.

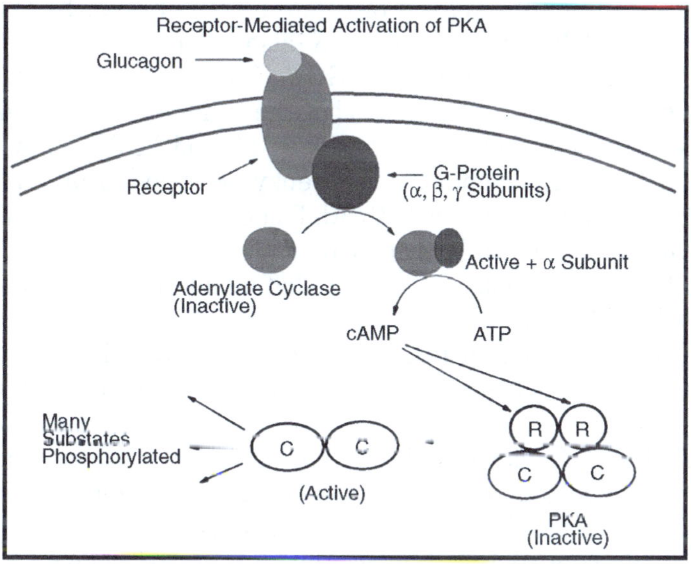

Fig. Activation of cAMP-dependent Protein Kinase (PKA).

In the preceding discussion we assumed that allosteric sites and catalytic sites were homogeneously present on every subunit of an allosteric enzyme. While this is often the case, there is another class of allosteric enzymes that are comprised of separate catalytic and regulatory subunits. The archetype of this class of enzymes is cAMP-dependent protein kinase (PKA), whose mechanism of activation.

The enzyme is tetrameric, containing two catalytic subunits and two regulatory subunits, and enzymatically inactive. When intracellular cAMP levels rise, one molecule of cAMP binds to each regulatory subunit, causing the tetramer to dissociate into one regulatory dimer and two catalytic monomers. In the dissociated form, the catalytic subunits are fully active; they catalyze the phosphorylation of a number of other enzymes, such as those involved in regulating glycogen metabolism. The regulatory subunits have no catalytic activity.

ENZYMES IN THE DIAGNOSIS OF PATHOLOGY

The measurement of the serum levels of numerous enzymes has been shown to be of diagnostic significance. This is because the presence of these enzymes in the serum indicates that tissue or cellular damage has occurred resulting in the release of intracellular components into the blood. Hence, when a physician indicates that he/she is going to assay for liver enzymes, the purpose is to ascertain the potential for liver cell damage.

Commonly assayed enzymes are the amino transferases: alanine transaminase, ALT (sometimes still referred to as serum glutamate-pyruvate aminotransferase, SGPT) and aspartate aminotransferase, AST (also referred to as serum glutamate-oxaloacetate aminotransferase, SGOT); lactate dehydrogenase, LDH; creatine kinase, CK (also called creatine phosphokinase, CPK); gamma-glutamyl transpeptidase, GGT. Other enzymes are assayed under a variety of different clinical situations but they will not be covered here.

The typical liver enzymes measured are AST and ALT. ALT is particularly diagnostic of liver involvement as this enzyme is found predominantly in hepatocytes. When assaying for both ALT and AST the ratio of the level of these two enzymes can also be diagnostic.

Normally in liver disease or damage that is not of viral origin the ratio of ALT/AST is less than 1.

However, with viral hepatitis the ALT/AST ratio will be greater than 1. Measurement of AST is useful not only for liver involvement but also for heart disease or damage.

The level of AST elevation in the serum is directly proportional to the number of cells involved as well as on the time following injury that the AST assay was performed. Following injury, levels of AST rise within 8 hours and peak 24-36 hours later. Within 3-7 days the level of AST should return to pre-injury levels, provided a continuous insult is not present or further injury occurs. Although measurement of AST is not, in and of itself, diagnostic for myocardial infarction, taken together with LDH and CK measurements the level of AST is useful for timing of the infarct.

The measurement of LDH is especially diagnostic for myocardial infarction because this enzyme exist in 5 closely related, but slightly different forms (isozymes).

The 5 types and their normal distribution and levels in non-disease/injury are listed below:

1. *LDH* 1: Found in heart and red-blood cells and is 17 per cent - 27 per cent of the normal serum total.
2. *LDH* 2: Found in heart and red-blood cells and is 27 per cent - 37 per cent of the normal serum total.
3. *LDH* 3: Found in a variety of organs and is 18 per cent - 25 per cent of the normal serum total.
4. *LDH* 4: Found in a variety of organs and is 3 per cent - 8 per cent of the normal serum total.
5. *LDH* 5: Found in liver and skeletal muscle and is 0 per cent - 5 per cent of the normal serum total.

Following a myocardial infarct the serum levels of LDH rise within 24-48 hours reaching a peak by 2-3 days and return to normal in 5-10 days. Especially diagnostic is a comparison of the LDH-1/LDH-2 ratio. Normally, this ration is less than 1. A reversal of this ration is referred to as a "flipped LDH". Following an acute myocardial infarct the flipped LDH ratio will appear in 12-24 hours and is definitely present by 48 hours in over 80% of patients. Also important is the fact that persons suffering chest pain due to angina only will not likely have altered LDH levels.

CPK is found primarily in heart and skeletal muscle as well as the brain. Therefore, measurement of serum CPK levels is a good diagnostic for injury to these tissues. The levels of CPK will rise within 6 hours of injury and peak by around 18 hours. If the injury is not persistent the level of CK returns to normal within 2-3 days.

Like LDH, there are tissue-specific isozymes of CPK and there designations:

- CPK3 (CPK-MM) is the predominant isozyme in muscle and is 100 per cent of the normal serum total.
- CPK2 (CPK-MB) accounts for about 35 per cent of the CPK activity in cardiac muscle, but less than 5 per cent in skeletal muscle and is 0 per cent of the normal serum total.
- CPK1 (CPK-BB) is the characteristic isozyme in brain and is in significant amounts in smooth muscle and is 0 per cent of the normal serum total.

Since most of the released CPK after a myocardial infarction is CPK-MB, an increased ratio of CPK-MB to total CPK may help in diagnosis of an acute infarction, but an increase of total CPK in itself may not. CPK-MB levels rise

3–6 hours after a myocardial infarct and peak 12–24 hours later if no further damage occurs and returns to normal 12–48 hours after the infarct.

OXIDOREDUCTASE ENZYMES

CARBONYL REDUCTIONS

The employment of whole-cell systems (*e.g.* bakers' yeast) or isolated enzymes for the reduction of carbonyl groups to the corresponding alcohols is well researched. Bakers' yeast has been used to reduce a wide range of simple ketones, β-ketoesters and cyclic diketones. The stereo-selectivity of the reduction process is predictable and very useful routes to homochiral anti-inflammatory steroids have been explored using this type of biotransformation. Other organisms, such as the fungus *Mortierella isabellina* can be used in place of bakers' yeast.

Partially purified dehydrogenase enzymes have also been utilised for the production of optically active secondary alcohols. During the purification of the protein the enzyme cofactor (nicotinamide adenine dinucleotide (phosphate) [NADPH]) is lost and so before the biotransformation is possible this cofactor must be re-coupled to the enzyme. The cofactors NADH and NADPH are expensive and cofactor recycling must be set up so that a less-than-stoichiometric amount of the dinucleotide can be employed. Cofactor recycling systems are now well defined thanks to the work of J. Bryan Jones and others, and are ready for use in non-specialist laboratories.

The ketone (25) has been reduced using *Mortierella* spp. and by 3ã,20ß-hydroxysteroid dehydrogenase/NADH. Thus the whole cell reduction gave an equimolar mixture of the 6*endo*-alcohol (26) and the 6*exo*-alcohol (27) (60 per cent yield of diastereoisomeric alcohols of high optical purity (Cotterill *et al.*, 1991) while the dehydrogenase catalysed an enantioselective reduction of the ketone (25) by NADH, forming the alcohol (26) in an optically pure state and leaving optically active ketone. The alcohol (26) has been converted into eldanolide, the pheromone of an agricultural pest, the sugar cane borer (Scheme 6). The availability of the two enantiomers of the 7,7-dimethylbicyclo[3.2.0]heptane ring system allows the synthesis of biologically interesting leukotriene-B4.

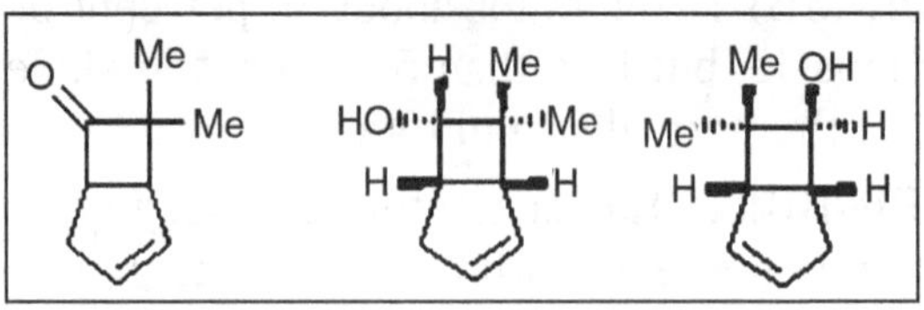

Note that one enantiomer of the ketone provides the C1-C6 section of the natural product (containing one chiral centre while the other enantiomer can provide the C7-C20 portion (containing the second chiral centre) (Scheme 6) (Cotterill *et al.*, 1990; Dormán *et al.*, 1989).

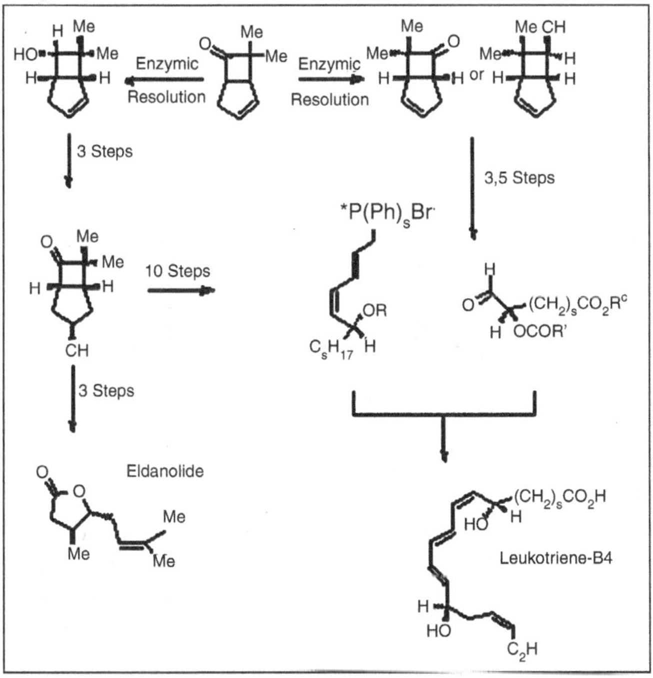

Reduction of Aromatic Nitro Groups

A less well studied reduction process is the yeast mediated reduction of aromatic nitro groups to the corresponding anilines.

We have been particularly interested in the opportunities for regioselective reduction of dinitroarenes.

In certain cases it is possible to achieve very high selectivity (*e.g.* reduction of the dinitroquinoline (28) to a single regioisomer of the product (29). On the basis of studying a large number of substrates we have formulated a series of rules to predict the outcome of the reduction.

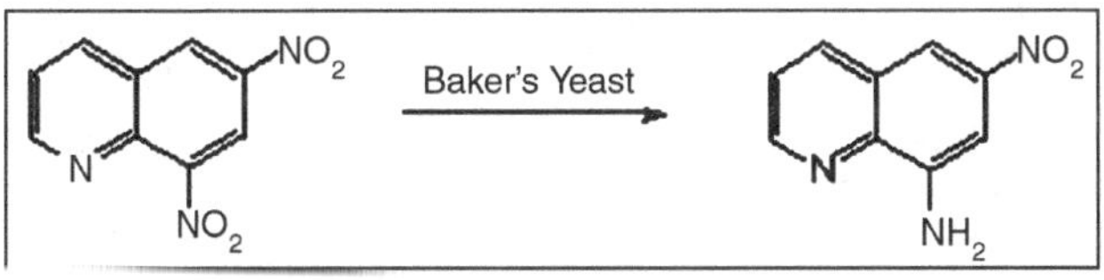

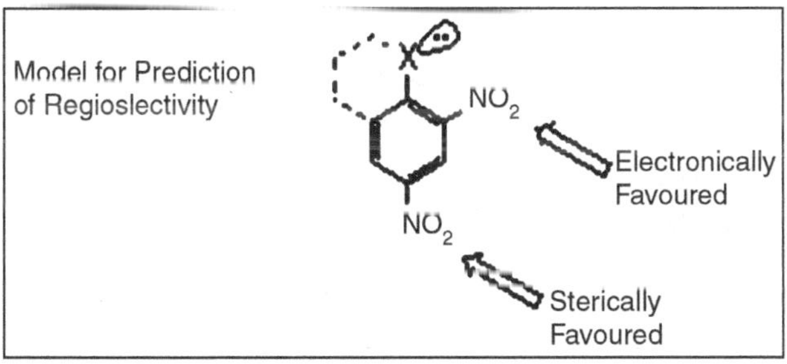

Oxidations

Oxidation of aromatic and alicyclic molecules using monooxygenase and dioxygenase enzymes has attracted a considerable amount of attention.

Dioxygenases

The oxidation of benzene to cyclohexa-3,5-diene-1,2-*cis*-diol is a reaction catalysed by *Pseudomonas putida* and substantial amounts of material can be obtained for further chemical studies.

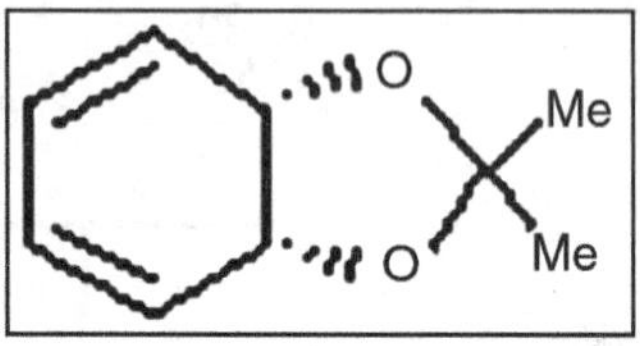

For example various cycloaddition reactions can be carried out on the acetonide (30), including Diels-Alder [4+2] reactions [to give, for example, the tricyclic compound (31)], [2+2] cycloaddition reactions [to give cyclobutanone derivatives such as (32)] and the more unusual [6+4] cycloaddition reactions.

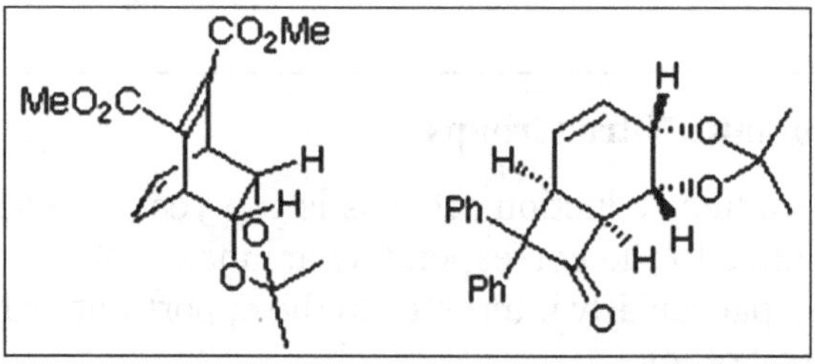

Not only is benzene biotransformed by *Pseudomonads* but so are many of its derivatives.

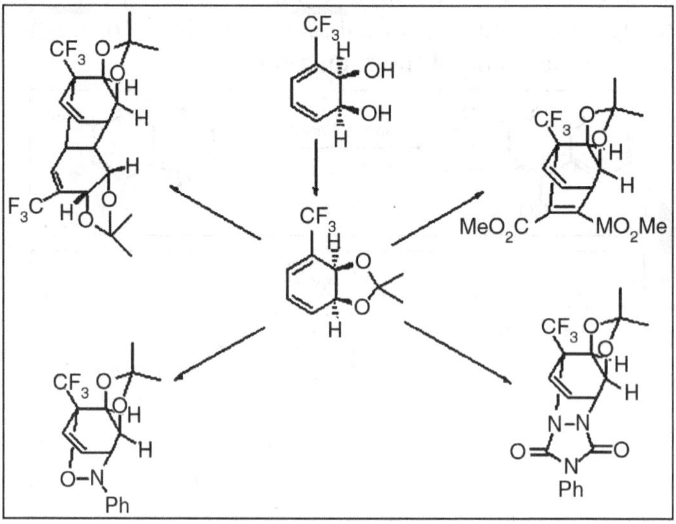

Often the latter compounds are obtained in an optically pure form and it is established (at least for some of the compounds) that these diols have the 1(*S*),2(*R*)-configuration. Trifluoromethylbenzene is biotransformed into the diol (33); the corresponding acetonide (34) can be converted into a wide range of derivatives through different chemical reactions. Interesting conversions involving 3-chlorocyclohexa-3,5-diene-1,2-diol and 3-methylcyclohexa-3,5-diene-1,2-diol have also been described.

Monooxygenases

Another oxidation reaction that is gaining an increasing amount of interest is the enzyme-catalysed conversion of a cyclic ketone into a lactone (Walsh and Chen, 1988).

We have shown that (+/-)-7-*anti*-fluoro-5*endo*-bromobicyclo[2.2.1]heptan-2-one is oxidized in an enantioselective manner using the *Acinetobacter* NCIMB 9871 to give optically pure lactone (35) (30 per cent) and recovered optically active ketone (36) (30 per cent). Interestingly oxidation of the ketone (36) with the chemical oxidant, *meta*-chloroperoxybenzoic acid gave the 3-oxabicyclooctan-2-one (37) as the major product. The latter compound was converted into the carbocyclic nucleoside (38), a compound possessing anti-HIV activity.

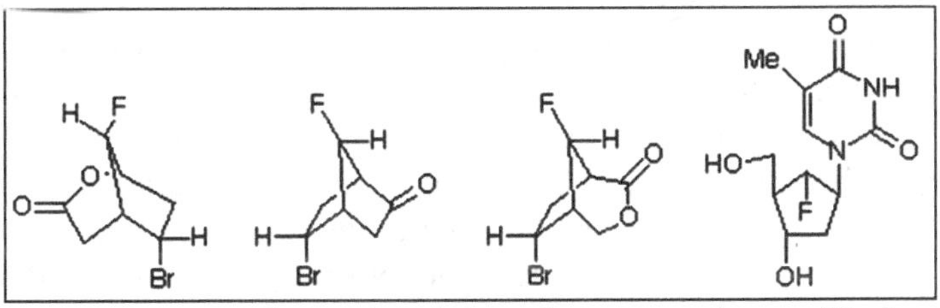

The monooxygenase from the *Acinetobacter* can be purified and Baeyer-Villiger oxidations can be performed using the enzyme in association with its cofactor NADPH. We have found that it is possible to convert a secondary alcohol into the corresponding lactone *via* the corresponding ketone by using a coupled enzyme system, that is *Thermoanaerobium brockii* dehydrogenase and *A. calcoaceticus* monooxygenase with *in situ* recycling of the cofactor. More recently a monooxygenase enzyme capable of using NADH as cofactor has been found, enabling cofactor recycling to be accomplished using formate dehydrogenase.

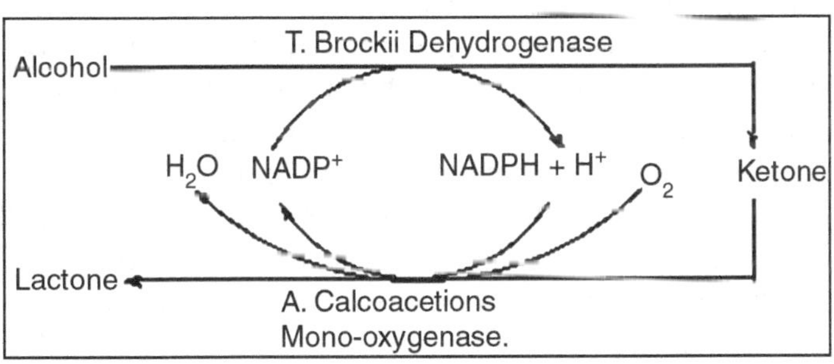

As adumbrated previously, immobilised soya bean lipoxygenase-1 efficiently converts linoleic acid into 13(*S*)-HODE (3) in 78 per cent yield in a few hours. The immobilised enzyme can be recovered and reused. The chemistry of 13-HODE has been investigated: for example the bicyclic compound (39) has been obtained by macro-lactonization and a Diels-Alder reaction on the biotransformation product.

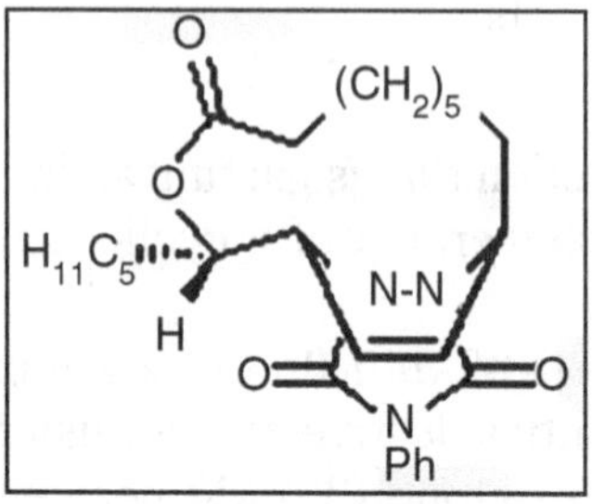

The ability to derivatize an organic molecule at a position distant from pre-existing functionality is an attractive goal for both organic chemists and microbiologists alike. Some progress has been made towards providing chemical methods for remote hydroxylation but, for the foreseeable future, microbiological methods seem to offer the most likely chances for success. The regioselective hydroxylation of steroids has been investigated in some depth, the formation of 11-hydroxyprogesterone from progesterone being one of the most noteworthy successes. The regio-selective (but not regiospecific) oxidation of cyclohexylcyclohexane to give the diol (40) using *Cunninghamella blakesleeana* shows that even the most hydrophobic materials can be biotransformed. Spiro-heterocyclic compounds can be derivatized, often in very high yield, to produce single compounds using *Beauveria sulfurescens*: for example the piperidine derivative (41) furnishes the alcohol (42) in 70 per cent isolated yield.

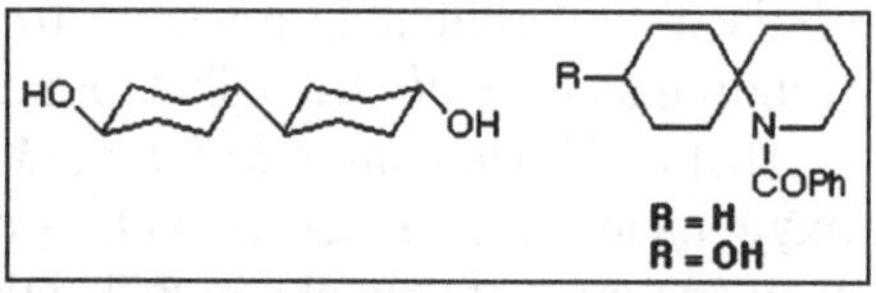

ENZYME NOMENCLATURE

All enzymes contain a protein backbone. In some enzymes this is the only component in the structure. However there are additional non-protein moieties usually present which may or may not participate in the catalytic activity of the enzyme. Covalently attached carbohydrate groups are commonly encountered structural features which often have no direct bearing on the catalytic activity, although they may well effect an enzyme's stability and solubility. Other factors often found are metal ions (cofactors) and low molecular weight organic molecules (coenzymes). These may be loosely or tightly bound by non-covalent or covalent forces. They are often important constituents contributing to both the activity and stability of the enzymes. This requirement for cofactors and coenzymes must be recognised if the enzymes are to be used efficiently and is particularly relevant in continuous processes where there may be a tendency for them to become separated from an enzyme's protein moiety.

Enzymes are classified according the report of a Nomenclature Committee appointed by the International Union of Biochemistry. This enzyme commission assigned each enzyme a recommended name and a 4-part distinguishing number. It should be appreciated that some alternative names remain in such common usage that they will be used, where appropriate, in this text.

The enzyme commission (EC) numbers divide enzymes into six main groups according to the type of reaction catalysed:

- Oxidoreductases which involve redox reactions in which hydrogen or oxygen atoms or electrons are transferred between molecules. This extensive class includes the dehydrogenases (hydride transfer), oxidases (electron transfer to molecular oxygen), oxygenases (oxygen transfer from molecular oxygen) and peroxidases (electron transfer to peroxide). For example: glucose oxidase.

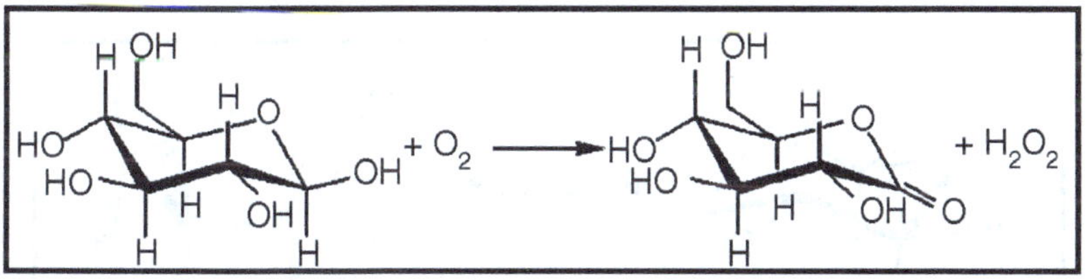

- β–D–glucose + oxygen→D-glucono–1,5-lactone + hydrogen peroxide
- Transferases which catalyse the transfer of an atom or group of atoms (*e.g.* acyl-, alkyl- and glycosyl-), between two molecules, but excluding such transfers as are classified in the other groups (*e.g.* oxidoreductases and hydrolases). For example: aspartate aminotransferase.

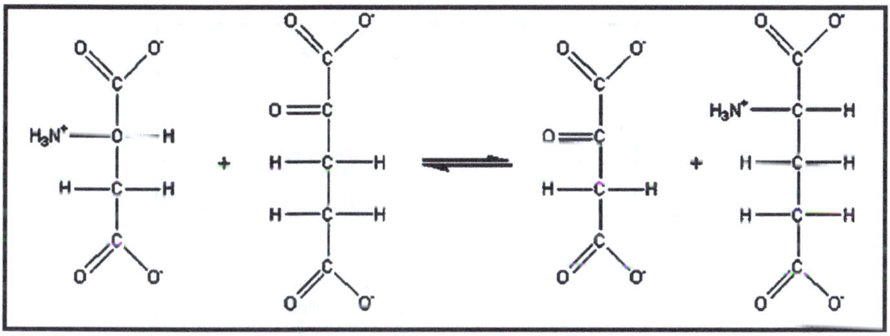

L-aspartate + 2–oxoglutarate ⇌ oxaloacetate + L–glutamate

- Hydrolases which involve hydrolytic reactions and their reversal.

This is presently the most commonly encountered class of enzymes within the field of enzyme technology and includes the esterases, glycosidases, lipases and proteases. For example: chymosin.

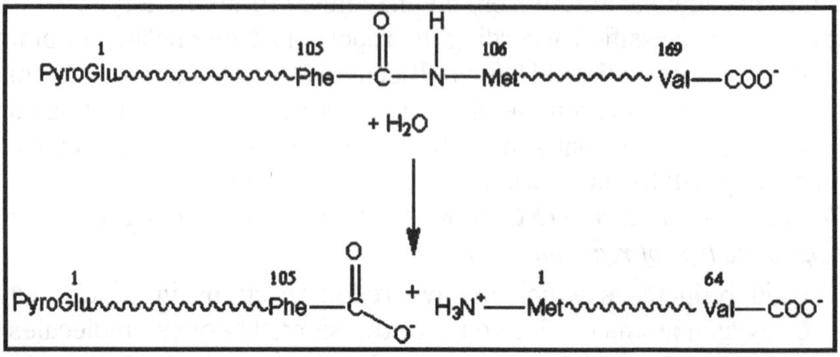

k-casein + water → *para*-k-casein + caseino macropeptide

- Isomerases which catalyse molecular isomerisations and includes the epimerases, racemases and intramolecular transferases. For example: xylose isomerase.

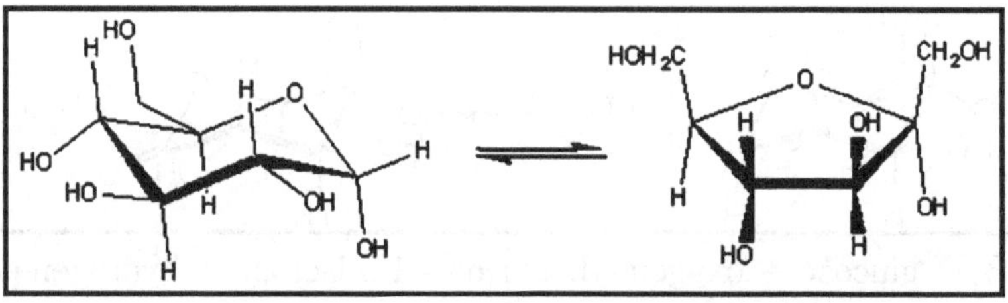

α-D-glucopyranose ⇌ α-D-fructofuranose

- Lyases which involve elimination reactions in which a group of atoms is removed from the substrate. This includes the aldolases, decarboxylases, dehydratases and some pectinases but does not include hydrolases. For example: histidine ammonia-lyase.

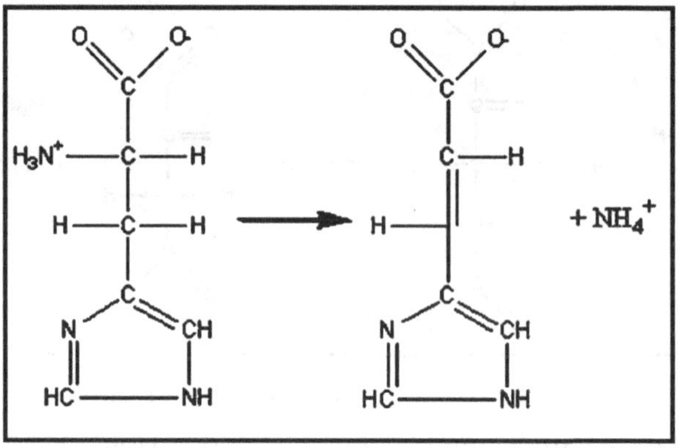

L-histidine → urocanate + ammonia

- Ligases, also known as synthetases, form a relatively small group of enzymes which involve the formation of a covalent bond joining two molecules together, coupled with the hydrolysis of a nucleoside triphosphate. For example: glutathione synthase (EC 6.3.2.3, systematic name,γ-L-glutamyl-L-cysteine:glycine ligase (ADP-forming); also called glutathione synthetase).

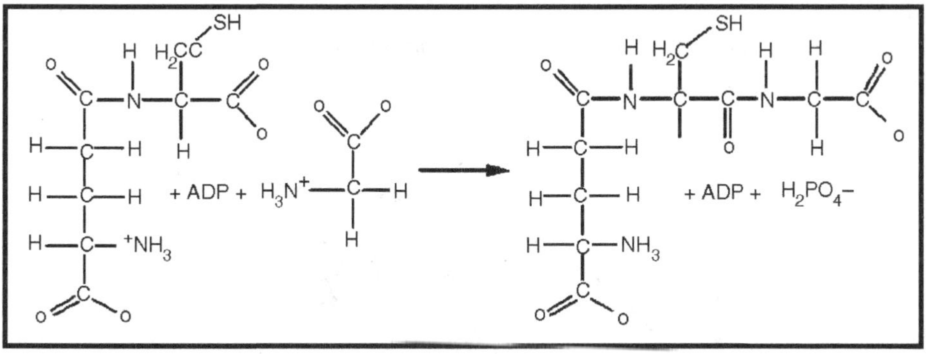

ATP + γ-L-glutamyl-L-cysteine + glycine → ADP + phosphate + glutathione.

CATALYZED REACTIONS

Examination of the change in reaction velocity as the reactant concentration is varied is one of the primary measurements in kinetic analysis. Returning to A → P, a plot of the reaction rate as a function of the concentration of A yields a straight line whose slope is *k* (Figure).

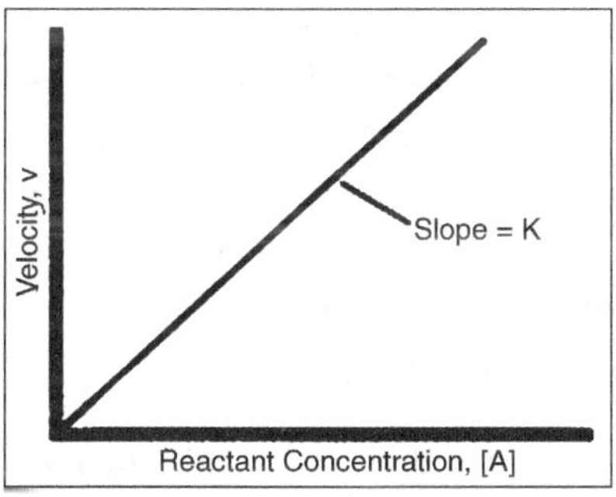

Fig. A Plot of v versus [A] for the Unimolecular Chemical Reaction, A →P, Yields a Straight Line having a Slope Equal to k

The more A that is available, the greater the rate of the reaction, v. Similar analyses of enzyme–catalyzed reactions involving only a single substrate yield remarkably different results (Figure).

At low concentrations of the substrate S, v is proportional to [S], as expected for a first–order reaction. However, v does not increase proportionally as [S] increases, but instead begins to level off. At high [S], v becomes virtually independent of [S] and approaches a maximal limit. The value of v at this limit is written V_{max}. Because rate is no longer dependent on [S] at these high concentrations, the enzyme–catalyzed reaction is now obeying zero–order kinetics; that is, the rate is independent of the reactant (substrate) concentration.

This behaviour is a saturation effect: when v shows no increase even though [S] is increased, the system is saturated with substrate. Such plots are called substrate saturation curves. The physical interpretation is that every enzyme molecule in the reaction mixture has its substrate–binding site occupied by S. Indeed, such curves were the initial clue that an enzyme interacts directly with its substrate by binding it.

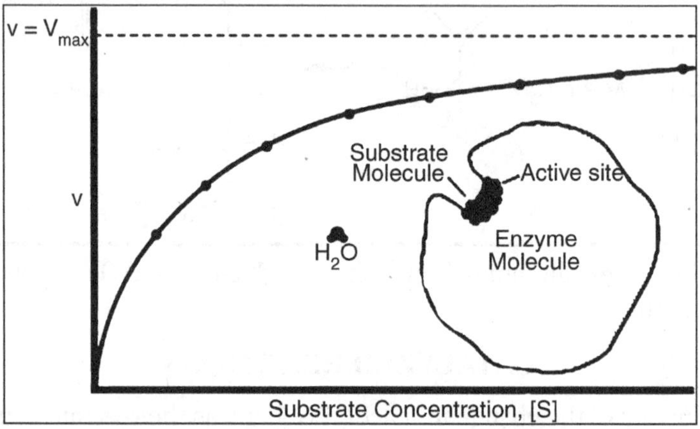

Fig. Substrate Saturation Curve for an Enzyme–catalyzed Reaction

The amount of enzyme is constant, and the velocity of the reaction is determined at various substrate concentrations. The reaction rate, v, as a function of [S] is described by a rectangular hyperbola. At very high [S], $v = V_{max}$. That is, the velocity is limited only by conditions (temperature, pH, ionic strength) and by the amount of enzyme present; v becomes independent of [S]. Such a condition is termed *zero–order kinetics*. Under zero–order conditions, velocity is directly dependent on [enzyme]. The H_2O molecule provides a rough guide to scale. The substrate is bound at the active site of the enzyme.

THE MICHAELIS – MENTEN EQUATION

Lenore Michaelis and Maud L. Menten proposed a general theory of enzyme action in 1913 consistent with observed enzyme kinetics. Their theory was based on the assumption that the enzyme, E, and its substrate, S, associate reversibly to form an enzyme–substrate complex, ES:

$$E + S \underset{k_{-1}}{\overset{k_1}{\rightleftharpoons}} ES$$

This association/dissociation is assumed to be a rapid equilibrium, and Ks is the *enzyme: substrate dissociation constant.* At equilibrium,

$$k_{-1}[ES] = k_1[E][S]$$

and

$$k_s = \frac{[E][S]}{[ES]} = \frac{k_1}{k_{-1}}$$

Product, P, is formed in a second step when ES breaks down to yield E + P.

$$E + S \underset{k_{-1}}{\overset{k_1}{\rightleftharpoons}} ES \xrightarrow{k_2} E + P$$

E is then free to interact with another molecule of S.

STEADY–STATE ASSUMPTION

The interpretations of Michaelis and Menten were refined and extended in 1925 by Briggs and Haldane, by assuming the concentration of the enzyme–substrate complex ES quickly reaches a constant value in such a dynamic system. That is, ES is formed as rapidly from E + S as it disappears by its two possible fates: dissociation to regenerate E + S, and reaction to form E + P. This assumption is termed thesteady–state assumption and is expressed as

$$\frac{d[ES]}{dt} = 0$$

That is, the change in concentration of ES with time, *t*, is 0. Figure illustrates the time course for formation of the ES complex and establishment of the steady–state condition.

INITIAL VELOCITY ASSUMPTION

One other simplification will be advantageous. Because enzymes accelerate the rate of the reverse reaction as well as the forward reaction, it would be helpful to ignore any back reaction by which E1P might form ES. The velocity of this back reaction would be given by $v = k_{-2}[E][P]$. However, if we observe only the *initial velocity* for the reaction immediately after E and S are mixed in the absence of P, the rate of any back reaction is negligible ecause its rate will be proportional to [P], and [P] is essentially 0. Given such simplification, we now analyse the system described by Equation in order to describe the initial velocity v as a function of [S] and amount of enzyme.

The total amount of enzyme is fixed and is given by the formula

Total enzyme,

$$[E_T] = [E] + [ES]$$

where [E] = free enzyme and [ES] = the amount of enzyme in the enzyme–substrate complex. From Equation, the rate of [ES] formation is,

$$v_f = k_1([ET] - [ES])[S]$$

where,

$$[ET] - [ES] = [E]$$

From Equation above, the rate of [ES] disappearance is,

$$v_d = k_{-1}[ES] + k_2[ES] = (k_{-1} + k_2)[ES]$$

At steady state,

$$d[ES]/dt = 0,$$

and therefore,

$$v_f = v_d.$$

So,

$$K_1([ET] - [ES])[S] = (k_{-1} + k_2)[ES]$$

Rearranging gives

$$\frac{([E_r]-[ES)[S]}{[ES]} = \frac{(k_{-1}^{!!} + k_2)}{k_1}$$

RELATIONSHIPS BETWEEN V_{MAX}, K_M, AND REACTION ORDER

The Michaelis–Menten equation describes a curve known from analytical geometry as a *rectangular hyperbola.*[1] In such curves, as [S] is increased, v approaches the limiting value, V_{max}, in an asymptotic fashion. V_{max} can be approximated experimentally from a substrate saturation curve (Figure), and K_m can be derived from $V_{max}/2$, so the two constants of the Michaelis–Menten equation can be obtained from plots of v versus [S].

Note, however, that actual estimation of V_{max}, and consequently K_m, is only approximate from such graphs. That is, according to Equation, in order to get v = 0.99 V_{max}, [S] must equal 99 K_m, a concentration that may be difficult to achieve in practice.

From Equation, when [S] >> K_m, then $v = V_{max}$. That is, v is no longer dependent on [S], so the reaction is obeying zero–order kinetics. Also, when [S] < K_m, then v » (Vmax/ K_m)[S]. That is, the rate, v, approximately follows a first–order rate equation, $v = k'[A]$, where

$$k' = V_{max}/K_m.$$

***K_m and V_{max}, once known explicitly, define the rate of the enzyme–catalyzed reaction, provided*:**

- The reaction involves only one substrate, *or* if the reaction is multisubstrate, the concentration of only one substrate is varied while the concentration of all other substrates is held constant.
- The reaction ES → E + P is irreversible, *or* the experiment is limited to observing only initial velocities where [P] = 0.
- $[S]_0 > [E_T]$ and $[E_T]$ is held constant.
- All other variables that might influence the rate of the reaction (temperature, pH, ionic strength, and so on) are constant.

ENZYMES AND LIFE PROCESSES

The living cell is the site of tremendous biochemical activity called metabolism. This is the process of chemical and physical change which goes on continually in the living organism.

Build-up of new tissue, replacement of old tissue, conversion of food to energy, disposal of waste materials, reproduction - all the activities that we characterize as "life." This building up and tearing down takes place in the face of an apparent paradox. The greatest majority of these biochemical reactions do not take place spontaneously.

The phenomenon of catalysis makes possible biochemical reactions necessary for all life processes. Catalysis is defined as the acceleration of a chemical reaction by some substance which itself undergoes no permanent chemical change. The catalysts of biochemical reactions are enzymes and are responsible for bringing about almost all of the chemical reactions in living organisms. Without enzymes, these reactions take place at a rate far too slow for the pace of metabolism.

The oxidation of a fatty acid to carbon dioxide and water is not a gentle process in a test tube - extremes of pH, high temperatures and corrosive chemicals are required. Yet in the body, such a reaction takes place smoothly and rapidly within a narrow range of pH and temperature. In the laboratory, the average protein must be boiled for about 24 hours in a 20 per cent HCl solution to achieve a complete breakdown. In the body, the breakdown takes place in four hours or less under conditions of mild physiological temperature and pH.

ENZYME CLASSIFICATIONS

Traditionally, enzymes were simply assigned names by the investigator who discovered the enzyme. As knowledge expanded, systems of enzyme classification became more comprehensive and complex. Currently enzymes are grouped into six functional classes by the International Union of Biochemists (I.U.B.).

Number	Classification	Biochemical Properties
1	Oxidoreductases	Act on many chemical groupings to add or remove hydrogen atoms.
2	Transferases	Transfer functional groups between donor and acceptor molecules. Kinases are specialized transferases that regulate metabolism by transferring phosphate from ATP to other molecules.
3	Hydrolases	Add water across a bond, hydrolyzing it.
4.	Lyases	Add water, ammonia or carbon dioxide across double bonds, or remove these elements to produce double bonds.

5	Isomerases	Carry out many kinds of isomerization: L to D isomerizations, mutase reactions (shifts of chemical groups) and others.
6	Ligases	Catalyze reactions in which two chemical groups are joined (or ligated) with the use of energy from ATP.

These rules give each enzyme a unique number. The I.U.B. system also specifies a textual name for each enzyme. The enzyme's name is comprised of the names of the substrate(s), the product(s) and the enzyme's functional class. Because many enzymes, such as alcohol dehydrogenase, are widely known in the scientific community by their common names, the change to I.U.B.-approved nomenclature has been slow. In everyday usage, most enzymes are still called by their common name.

Enzymes are also classified on the basis of their composition. Enzymes composed wholly of protein are known as simple enzymes in contrast to complex enzymes, which are composed of protein plus a relatively small organic molecule. Complex enzymes are also known as holoenzymes. In this terminology the protein component is known as the apoenzyme, while the non-protein component is known as the coenzyme or prosthetic group where prosthetic complex in which the small organic molecule is bound to the apoenzyme by covalent bonds; when the binding between the apoenzyme and non-protein components is non-covalent, the small organic molecule is called a coenzyme.

Many prosthetic groups and coenzymes are water-soluble derivatives of vitamins. It should be noted that the main clinical symptoms of dietary vitamin insufficiency generally arise from the malfunction of enzymes, which lack sufficient cofactors derived from vitamins to maintain homeostasis.

The non-protein component of an enzyme may be as simple as a metal ion or as complex as a small non-protein organic molecule. Enzymes that require a metal in their composition are known as metalloenzymes if they bind and retain their metal atom(s) under all conditions with very high affinity. Those which have a lower affinity for metal ion, but still require the metal ion for activity, are known as metal-activated enzymes.

TYPES OF ENZYMATIC REACTIONS

When looking at enzyme names such as 'transketolase' or 'phosphorylase', you will note that these names don't tell you exactly what reactions the enzymes may catalyze. A complete description should mention the coenzymes required, the substrates and the particular bonds in the substrates that are being severed or created. A nomenclature that meets these criteria has been developed by the Enzyme Commission of the IUBMB (International Union of Biochemistry and Molecular Biology). In the IUBMB nomenclature, the enzyme transketolase bears the formidable name:

Sedoheptulose-7-phosphate: D-glyceraldehyde-3-phosphate glycolaldehydetransferase. Such names, of course, are rather lengthy, and their use is not very widespread. To make the tasks of tracking and bookkeeping more manageable, these names are supplemented with numeric codes.

In the IUBMB scheme, enzymes are put into one out of six classes according to the reactions they catalyze.

These classes are:

- *Oxidoreductases*. These catalyze redox reactions, frequently involving one of the coenzymes NAD^+, $NADP^+$, or FAD.
- *Transferases*. These bring about the transfer of functional groups – *e.g.*, phosphate groups from ATP to another metabolite, which activates the latter and sets it up for subsequent reaction steps.
- *Hydrolases*. These catalyze hydrolysis reactions – *e.g.*, such as those involved in digestion of foodstuffs.
- *Lyases* – these effect elimination reactions that result in the formation of double bonds.
- *Isomerases*. These facilitate the interconversion of isomers. We will meet two examples as soon as we get into glycolysis.
- Ligases, which form new covalent bonds at the expense of ATP hydrolysis.

Of course, within each of these main classes, there are subclasses and sub-sub classes that correspond to details of substrates and mechanisms of the enzyme reactions. Each individual enzyme activity is assigned an individual number within a sub-sub class, so that we wind up with a four-figure designation, which is preceded by the letters 'EC' (Enzyme Commission). One good thing about this classification is that it is rarely used – the 'recommended names', which most of the time happen to be the traditional ones, are used instead.

The other good thing is that it has a sound appreciation of priorities. The single most important enzyme in student lifestyle – namely, alcohol dehydrogenase (or, as IUBMB puts it, alcohol:NAD oxidoreductase). This beneficial enzyme, residing in the liver, degrades ethanol, and without it you would be drunk all the time!

Enzyme Kinetics: Basic Enzyme Reactions

Enzymes are catalysts and increase the speed of a chemical reaction without themselves undergoing any permanent chemical change. They are neither used up in the reaction nor do they appear as reaction products.

The basic enzymatic reaction can be represented as follows where E represents the enzyme catalyzing the reaction, S the substrate, the substance being changed, and P the product of the reaction.

Enzyme Kinetics: Energy Levels

Chemists have known for almost a century that for most chemical reactions to proceed, some form of energy is needed. They have termed this quantity of energy, "the energy of activation." It is the magnitude of the activation energy which determines just how fast the reaction will proceed. It is believed that enzymes lower the activation energy for the reaction they are catalyzing. Figure illustrates this concept.

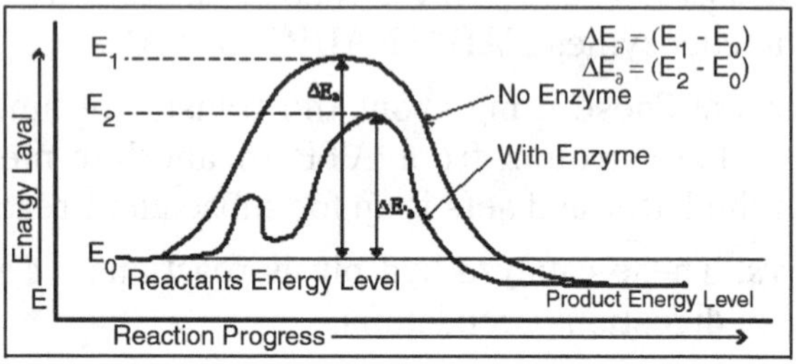

Fig. Energy of Activation Δ_∂ E is less than Δ_∂ E due to the Effect of the Enzyme on the Substrate.

The enzyme is thought to reduce the "path" of the reaction. This shortened path would require less energy for each molecule of substrate converted to product. Given a total amount of available energy, more molecules of substrate would be converted when the enzyme is present (the shortened "path") than when it is absent. Hence, the reaction is said to go faster in a given period of time.

CHEMICAL NATURE OF ENZYMES

All known enzymes are proteins. They are high molecular weight compounds made up principally of chains of amino acids linked together by peptide bonds.

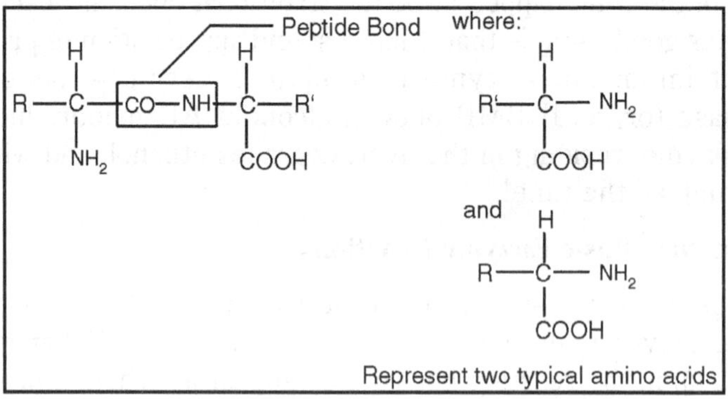

Fig. Typical Protein Structure - Two Amino Acids Joined by a Peptide Bond.

Enzymes can be denatured and precipitated with salts, solvents and other reagents. They have molecular weights ranging from 10,000 to 2,000,000. Many enzymes require the presence of other compounds - cofactors - before their catalytic activity can be exerted.

This entire active complex is referred to as the holoenzyme; *i.e.*, apoenzyme (protein portion) plus the cofactor (coenzyme, prosthetic group or metal-ion-activator) is called the holoenzyme.

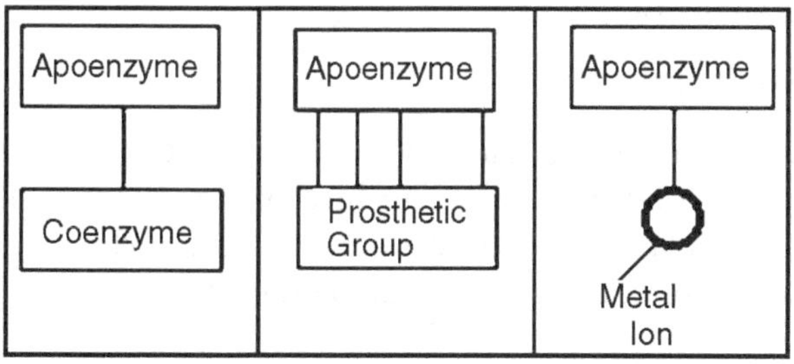

Fig. Holoenzymes- Apoenzymes Plus Various Types of Cofactors.

Apoenzyme + Cofactor = Holoenzyme

According to Holum, the cofactor may be:

- *A coenzyme*: A non-protein organic substance which is dialyzable, thermostable and loosely attached to the protein part.
- *A prosthetic group*: An organic substance which is dialyzable and thermostable which is firmly attached to the protein or apoenzyme portion.
- *A metal-ion-activator*: These include K^+, Fe^{++}, Fe^{+++}, Cu^{++}, Co^{++}, Zn^{++}, Mn^{++}, Mg^{++}, Ca^{++}, and Mo^{+++}.

Specificity of Enzymes

One of the properties of enzymes that makes them so important as diagnostic and research tools is the specificity they exhibit relative to the reactions they catalyze.

A few enzymes exhibit absolute specificity; that is, they will catalyze only one particular reaction. Other enzymes will be specific for a particular type of chemical bond or functional group.

In general, there are four distinct types of specificity:

1. *Absolute specificity*: The enzyme will catalyze only one reaction.
2. *Group specificity*: The enzyme will act only on molecules that have specific functional groups, such as amino, phosphate and methyl groups.
3. *Linkage specificity*: The enzyme will act on a particular type of chemical bond regardless of the rest of the molecular structure.
4. *Stereochemical specificity*: The enzyme will act on a particular steric or optical isomer.

Though enzymes exhibit great degrees of specificity, cofactors may serve many apoenzymes. For example, nicotinamide adenine dinucleotide (NAD) is a coenzyme for a great number of dehydrogenase reactions in which it acts as a hydrogen acceptor.

Among them are the alcohol dehydrogenase, malate dehydrogenase and lactate dehydrogenase reactions.

Naming and Classification

Except for some of the original enzymes such as pepsin, rennin, and trypsin, most enzyme names end in "ase". The International Union of Biochemistry (I.U.B.) initiated standards of enzyme nomenclature which recommend that enzyme names indicate both the substrate acted upon and the type of reaction catalyzed.

Under this system, the enzyme uricase is called urate: O_2 oxidoreductase, while the enzyme glutamic oxaloacetic transaminase (GOT) is called L-aspartate: 2-oxoglutarate aminotransferase.

Enzymes can be classified by the kind of chemical reaction catalyzed:

- Addition or removal of water
  - *Hydrolases*: These include esterases, carbohydrases, nucleases, deaminases, amidases, and proteases
  - Hydrases such as fumarase, enolase, aconitase and carbonic anhydrase
- Transfer of electrons
  - Oxidases
  - Dehydrogenases
- Transfer of a radical
  - *Transglycosidases*: Of monosaccharides
  - *Transphosphorylases and phosphomutases*: Of a phosphate group
  - *Transaminases*: Of amino group
  - *Transmethylases*: Of a methyl group
  - *Transacetylases*: Of an acetyl group
- Splitting or forming a C-C bond
  - Desmolases
- Changing geometry or structure of a molecule
  - Isomerases
- Joining two molecules through hydrolysis of pyrophosphate bond in ATP or other triphosphate
  - Ligases.

HOW ENZYMES WORK

For two molecules to react they must collide with one another. They must collide in the right direction (orientation) and with sufficient energy. Sufficient energy means that between them they have enough energy to overcome the energy barrier to reaction. This is called the activation energy. Enzymes have an active site. This is part of the molecule that has just the right shape and functional groups to bind to one of the reacting molecules. The reacting molecule that binds to the enzyme is called the substrate. An enzyme-catalysed reaction takes a different 'route'. The enzyme and substrate form a reaction intermediate. Its formation has a lower activation energy than the reaction between reactants without a catalyst.

A simplified picture:

Route A reactant 1 + reactant 2 → product

Route B reactant 1 + enzyme → intermediate

intermediate + reactant 2 → product + enzyme

So the enzyme is used to form a reaction intermediate, but when this reacts with another reactant the enzyme reforms.

LOCK AND KEY HYPOTHESIS

This is the simplest model to represent how an enzyme works. The substrate simply fits into the active site to form a reaction intermediate.

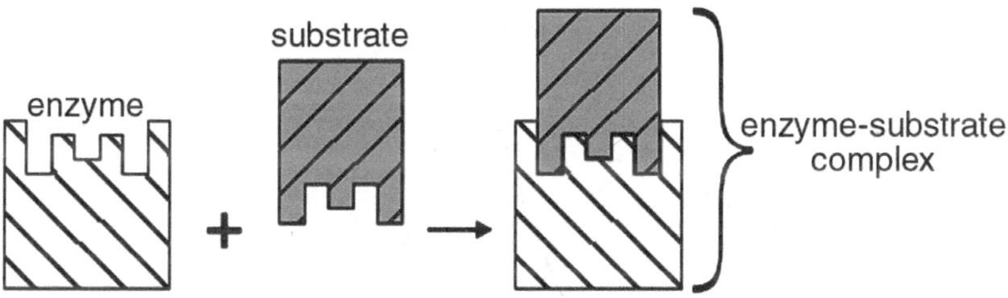

Induced Fit Hypothesis

In this model the enzyme molecule changes shape as the substrate molecules gets close. The change in shape is 'induced' by the approaching substrate molecule. This more sophisticated model relies on the fact that molecules are flexible because single covalent bonds are free to rotate.

Factors Affecting Catalytic Activity of Enzymes

Temperature

As the temperature rises, reacting molecules have more and more kinetic energy. This increases the chances of a successful collision and so the rate increases. There is a certain temperature at which an enzyme's catalytic activity is at its greatest. This optimal temperature is usually around human body temperature (37.5 °C) for the enzymes in human cells.

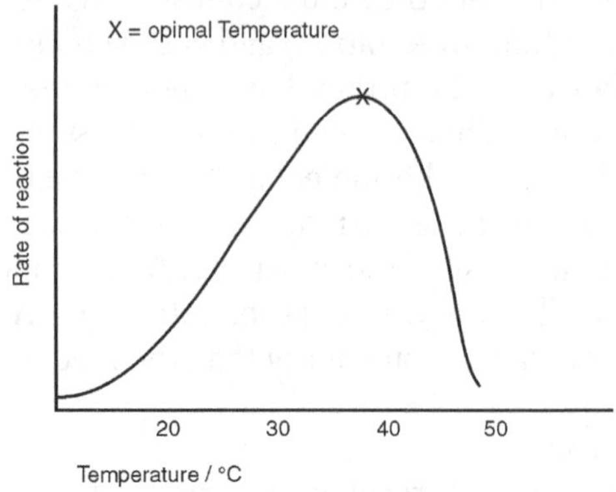

pH

Below this temperature the enzyme structure begins to break down (denature) since at higher temperatures intra-and intermolecular bonds are broken as the enzyme molecules gain even more kinetic energy. Each enzyme works within quite a small pH range. There is a pH at which its activity is greatest (the optimal pH). This is because changes in pH can make and break intra-and intermolecular bonds, changing the shape of the enzyme and, therefore, its effectiveness.

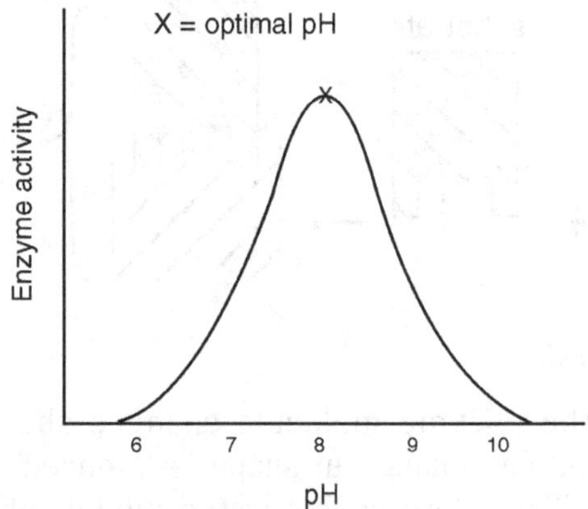

Concentration of Enzyme and Substrate

The rate of an enzyme-catalysed reaction depends on the concentrations of enzyme and substrate. As the concentration of either is increased the rate of reaction increases. For a given enzyme concentration, the rate of reaction increases with increasing substrate concentration up to a point, above which any further increase in substrate concentration produces no significant change in reaction rate. This is because the active sites of the enzyme molecules at any given moment are virtually saturated with substrate. The enzyme/substrate complex has to dissociate before the active sites are free to accommodate more substrate Provided that the substrate concentration is high and that temperature and pH are kept constant, the rate of reaction is proportional to the enzyme concentration.

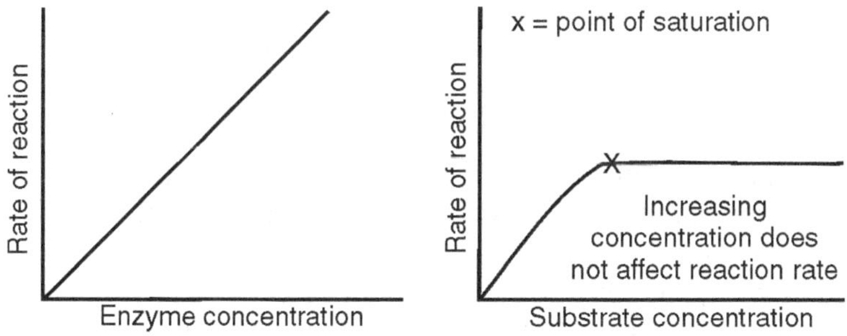

Inhibition of Enzyme activity

Some substances reduce or even stop the catalytic activity of enzymes in biochemical reactions. They block or distort the active site.

These chemicals are called inhibitors, because they inhibit reaction. Inhibitors that occupy the active site and prevent a substrate molecule from binding to the enzyme are said to be active site-directed (or competitive, as they 'compete' with the substrate for the active site). Inhibitors that attach to other parts of the enzyme molecule, perhaps distorting its shape, are said to be non-active site-directed (or non- competitive).

Immobilized Enzymes

Enzymes are widely used commercially, for example in the detergent, food and brewing industries.

Protease enzymes are used in 'biological' washing powders to speed up the breakdown of proteins in stains like blood and egg. Pectinase is used to produce and clarify fruit juices.

Problems using enzymes commercially include:

- They are water soluble which makes them hard to recover
- Some products can inhibit the enzyme activity (feedback inhibition) Enzymes can be immobilized by fixing them to a solid surface. This has a number of commercial advantages:
- The enzyme is easily removed
- The enzyme can be packed into columns and used over a long period
- Speedy separation of products reduces feedback inhibition
- Thermal stability is increased allowing higher temperatures to be used
- Higher operating temperatures increase rate of reaction

There are four principal methods of immobilization currently in use:

- Covalent bonding to a solid support
- Adsorption onto an insoluble substance
- Entrapment within a gel
- Encapsulation behind a selectively permeable membrane

THE GENERAL PROPERTIES OF ENZYMES

All enz)niies are colloids, like the native proteins of the seralbumin type. Like these, they cannot be dialysed through parchment, and are irrevocably changed on heating in aqueous solution to 60° to 70° C. The albumin is coagulated by heat, and this coagulum cannot be dissolved without conversion into peptones; the enzymes, however, disappear entirely on heating.

It is possible that they are coagulated and become insoluble; it is also possible that they are chemically changed and decomposed. The only sign of the presence of enzymes is their action, and we see that the hydrolysing or oxidising power of a solution is lost by heating, and cannot be restored. Pfeflfer was able to show that the enzymes do not volatiUze with the vapor of water and thus do not leave the solution.

All enzymes are destroyed by heat. Most of them are more or less slowly impaired at a low temperature 37° to 40° C. – ^ and many undergo a slow loss of power, at room temperature, and even at zero. Zymase and the glycolytic enzyme of muscles are destroyed in the frozen state in two to four days. ^ Enzymes are also broken down by many chemical processes, by strong acids and alkalies, by alcohol, etc. Pepsin loses its enzymotic properties by contact with the weakest alkali, and many oxidising enzymes by acid. It is possible that this alteration is the same as that which takes place at a slower rate on heating. Bayliss* supposes that the alteration at low temperature is reversible. He cites as an example, that toxins may become inactive toxoids while retaining their chemical character.

This supposition was re- cently supported in a very remarkable manner by Pawlow.' It has always been believed that the pepsin is completely destroyed by alkali; Pawlow and Tichomirow demonstrat- ed that this is the case only if the alkaline solution is im mediately acidified; the pepsin cannot then be detected. If, however, the gastric juice be made alkaline, then neutralised, and allowed to stand for some time before acidulating it, the pepsin does not disappear, but acts almost as strongly as before.

This is the first case of restitution of an enzyme after having lost its vigour. Melt- zer has pointed out that this slow loss of power is greatly accelerated on shaking solutions of enzymes. The de- struction through shaking takes place even at low tem- peratures, but proceeds more rapidly at the temperature of the body.

To-day every investigation of enzymes is retarded by the same difficulties as those attending the chemical study of proteins. The true proteins are like enzymes so far as sensitiveness towards temperature and towards acid and alkali are concerned. Knowledge of the chemistry of the proteins stagnated for a long time, because the methods of dealing with colloids were undeveloped.

It made very rapid progress as soon as it was found possible to work with strong hydrochloric acid, barium hydroxide, and phosphotungstic acid. I have cited the methods of purifying enzymes, salting out, and precipitating by weak acid or alcohol, avoiding heat. By such methods we are unable to separate compounds which can be chemically defined. It is to be hoped that Pawlow's observation may become the beginning of a really chemical explora- tion of enzymes.

' We cannot detect or estimate the enzymes themselves. We can only observe their action. For example, we can estimate the quantity of cane sugar before and after the ac- tion of invertin; we can also determine the quantity of glucose and levulose, the products of the decomposition of the cane sugar. Further, we can estimate the quantity of the coagulable proteins before the action, and after five, ten, and fifteen minutes, or the quantity of the non-coagulable peptone; but we cannot directly estimate the invertin or pepsin.

Therefore everything that disturbs the action of the enzyme seems to be a diminution, while everytlnng that improves the action seems to be an increase, of the quantity of the enzyme, although possibly only due to differences in the conditions of its action. rule, generally acknowledged in chemistry, that for deternjining the physical properties of a substance - ^the molecular weight, the melting-point, the osmotic pressure, the rotatory power, the electrical conductivity, the ioniza- tion - ^the substance must be wholly purified.

We know that in enyzme solutions the weight of enzymes is much smaller than the weight of the impurities; it is clear that in such a solution we cannot define the law of action of these bodies. If we read the books or papers written on enzymes by chemists, we find that the law of Schultz states that the action of a solution of an enzyme increases, not in proportion to the quantity, but to the square root of the quantity of the contained enzyme; and the question is also, debated, whether the enzyme is consumed during its action or not, etc.

The prevailing great interest in the enzymes has led authors to bring up questions which no body can answer at the present time. All we can do at present is to describe a number of the properties of enzymes which have been observed.

- Enzymes are colloids, and do not dialyse through parchment.
- Enzymes have an optimum temperature of action. It is well known that all chemical processes go faster with increasing temperature, and van 't Hoff enunciated the rule that the rapidity of most chemical processes is doubled or trebled when the temperature is raised io° C. The enzymes follow this rule, but only between o° and 40° C.

above 40° C, this power decreases rapidly. If we graphi cally represent the behaviour of the decomposition of canesugar by hydrochloric acid and by invertin, with increasing temperature, we find for the acid a curve like this, and for the enzyme a curve/^, *i.e.*, a curve with a dis- tinct maximum. It is possible that two processes are involved: the general increasing reaction with rise of temperature, and the destruction of the enzymes by heat. But this is not certain. In some enzymes we can see a slow acceleration of action at from io° to 30° C, and a much more rapid acceleration at from 30° to 37° C. The curve has a form difficult to explain as the summation of opposed effects. It is possible that we have, besides the general acceleration with rising temperature, a special adaptation of the enzymes to the body-temperature of the higher animals. The optimum of most of the enzymes examined lies between 30° and 40° C. But we find old statements that the optimum of the diastase of yeast lies at 50° C, or that the pepsin of fish has an optimum between 2° and 5° C. These statements do not hold good, as the tissues of cold-blooded animals contain proteins coagulat- ing spontaneously at 30° to 40° C; if these proteins are precipitated, pepsin is precipitated by adsorption, and the extract becomes accordingly poorer without change of the pepsin itself. In the case of yeast, the better solubility at higher temperatures can explain the obser\'ation. Until we can work with pure enzymes, we shall have great difficulties in distinguishing a change in solubility from a true change in the enzyme.

- Enzymes and Reaction. Most enzymes are exactly adapted to the reaction of their "milieu" - that is to say, of the solution in which they occur in nature. Thus pepsin acts best in combination with acid, with hydrogen- ions. It does not digest protein, but only the chloride of the protein, and the products of the enzyme are the chlorides of peptones. The quantity of acid which is bound on the proteins and peptones, and which changes the free compounds into chlorides, is not suflScient for the action of pepsin. This was explained by Leo.^ Pepsin needs an excess of acid for its action, that is to say, it needs free hydrogen-ions. When we study dissolved protein, we cannot see this, because the hydrochloric- proteins split oflE acid by hydrolysis. But if we use a solid protein, for instance fibrin, we find that we must add so much hydrochloric acid, that the water, in which the swollen fibrin lies, still contains acid. The natural gastric juice of the dog or of man contains one-half or more per cent of hydrochloric acid, but the juice is diluted by food to 0.35 per cent; it is then i/io normal hydro- chloric acid, and it has been pointed out by Briicke^ and Pawlow, that the optimum of pepsin lies at this lower concentration. The hydrochloric acid can be replaced by another acid according to the available hydrogen-ions, but it seems that the hydrochloric acid has an exceptional value, which is greater than that of other acids of the same concentration and ionisation.

- The necessity of co-operation between pepsin and hydrochloric acid has caused many errors in the chemistry of enzymes. The peptones produced by pepsin are basic, and neutralise the acids; the hydrochloric-peptones are hydrolysed salts, and the quantity of acid set free by hydrolysis depends upon the concentration and upon the presence of other neutral salts. All these influences appear to us as a diminution or an increase of pepsin. Furthermore, we know two proteins, the myosin of muscles or syntonin, and the fibrin of the blood, which swell in acids and can be attacked by pepsin only in the swollen state. Sodium chloride and other neutral salts check this swelling, and appear as opposed to pepsin; as '^antiferments." Pepsin is least suitable as an example
- For pointing out the laws of enzymotic action.Un- fortunately, however, it is used most frequently. The relations of trypsin to the reaction have often been in- vestigated by earlier authors, but they did not always distinguish between the speed of action of the dissolved trypsin and the ease of dissolving trypsin from the gland.

The enzyme has two actions: it dissolves the protein, and it splits up proteins and peptones into simpler products. It also seems to have two different optima. It dissolves proteins when the reaction is distinctly alkaline; and it decomposes peptones best at the approximately neutral or slightly alkaline reaction which prevails in the intestine. This nearly neutral reaction is best also for the enzymes of the saliva, the intestine, and the tissues. For a long time there has been a division of opinion about the re- action of the contents of the intestine and the tissues.

The dispute arose solely because of the deficient knowledge at the time of the qualities of the indicators used in the study of the reaction. Now we know that the reaction of both tissues and contents of the intestine resembles that of a weak solution of alkali saturated with an excess of carbonic acid. Such a solution gives the optimum for the enzymes of the saliva and intestine, and for the oxidising enzymes of the tissues. The tissue enzymes are destroyed by an abnormal reaction, but the enzymes of the intestine show a greater or less resistance towards difi^erences in reaction, a remarkable adaptation to the conditions of their environment, because the reaction in the upper duodenum changes often and suddenly.

Zymase, the metabolic enzyme of the yeast, acts only, or best, in the presence of phosphates; of sodium mono- or diphosphate. Perhaps this is an example of a true activator, a matter which will be dealt with later on, but more probably the power of the phosphoric acid to hinder marked changes of reaction is the cause of the favourable influence of the phosphates, and this is therefore a fur- ther exemplification of the importance of reaction upon ferments.

- *Chemical Properties*: Enzymes combine both with their substrate and their dissociation products. The specificity of the enzymes mentioned before suggests to us, that the individual enzyiaes must have chemical relations with this substrate. We have no reason to think that all enzymes belong to the same class of chemical compounds. It has already been shown that the enzymes are accompanied as a rule by proteins and nucleic acids.

The difficulties of separating enzymes and proteins have led physiologists for a long time to regard enzymes as protein-like bodies, though Briicke * and others ^ were successful years ago in freeing enzymes from all traces of proteins. Perhaps steapsin is a fat-like compound, and sucroclastic enzymes sugar-like compounds, and perhaps they belong to a wholly different class. We know but little to-day about the chemical characteristics of enz)anes.

Enzymes are salted out by such neutral salts as are capable of readily salting out proteins, uric acid, and many dye-compounds, and particularly by anmionium sulphate. It seems that the individual enzymes are precipitated at diflferent concentrations of ammonium sulphate - ^aldehydase and erepsin, when the solution contains sixty per cent, and trypsin and invertin at complete saturation.

These diflferences can be used practically for purifying and separating enzymes, but they throw no light on their chemical nature. Enzymes are precipitated by alcohol, and the individual enzymes apparently by different concentrations of alcohol. En- zymes are precipitated, further, by heavy metals like uranium, and they are precipitated by tannin and by phosphotiingstic acid. The enzymes can be dissolved again immediately after precipitation, but on standing they readily lose their solubility. Colour tests for enzymes are not known.

It is of practical importance that enz)mies are neither precipitated nor destroyed by many antiseptic substances, *i.e.*, by substances that kill and destroy living cells or check their growth. Chloroform, toluene, thymol, and ether, which dissolve the lipoid substances in cells and bacteria, and therefore destroy structures necessary to life, exert no influence on enzymes, because the enz)anes are completely soluble in water, and do not yield any lipoids. It is very difficult to get extracts of living tissues, which cannot be heated or boiled, free from bacteria; so far as concerns the liver and pancreas, which yield bacteria in life, it is impossible.

During life, growth of micro-organisms is prevented or controlled in organs, but after death and loss of structure, bacteria grow in the fluids rich in proteins, carbohydrates, and suitable salts. These bacteria split the proteins, carbohydrates, and fats, as do the hydrolytic ferments, and give rise, like zymase, to carbon dioxide. Every one who has worked with ferments knows how easily we can be deceived when led to suppose ferments to be present when in fact we are dealing with bacteria.

We distin- guish between the two when we allow ferments to work for only two or three hours, because the action of ferments proceeds very quickly, whilst the growth of bacteria, even under the most favourable conditions, takes much more time.

But during extraction and purification, and if we extend experiments for a long time, it is absolutely necessary to protect enzyme solutions against bacteria by means of antiseptic substances.

Amongst these chloroform or toluene have been most used within the last few years, and for the study of the enzymes of the alimentary canal crowded with bacteria, I suggest using both substances simultaneously. Thymol and ether have no efficiency.

I think that all observations on enzymes, made without any addition of antiseptics, must be regarded with doubt and distrust. Other antiseptics, which coagulate proteins, such as fluorides, potassium meta-arsenite, and mercuric chloride, seem to have an injurious effect on the action of the enzymes. But no evidence has been brought forward showing whether ferments are precipitated and destroyed by these compounds, or whether the proteins, on coagulat- ing, absorb and withdraw the enzymes. Buchner ^ has observed that zymase does not lose power in the presence of potassium meta-arsenite or ammonium fluoride in small amount, and that the proteolytic ferments of leucocytes are not harmed by corrosive sublimate.

But these solu- tions, zymase and the extracts of leucocytes, are rich in protein, and the harmful influence of the mercury or the arsenite is checked by their union with protein. Purified enzyme solutions have not yet been studied with regard to their sensitiveness to these antiseptics, but for most practical purposes the lipoid antiseptics are thoroughly satisfactory.

BIOLOGICAL CATALYSTS OF ENZYMES

This means that they speed up the chemical reactions in living things. Without enzymes, our guts would take weeks and weeks to digest our food, our muscles, nerves and bones would not work properly and so on - we would not be living!

A catalyst is any substance which makes a chemical reaction go faster, without itself being changed. A catalyst can be used over and over again in a chemical reaction: it does not get used up. Enzymes are very much the same except that they are easily denatured (destroyed: but do NOT use this word since the protein molecule is not broken down into amino-acids, it just loses it shape and will not work any more) by heat. Our enzymes work best at body temperature. Our enzymes also have to have the correct pH. pH and heavy metal ions. Unlike ordinary catalysts, they are specific to one chemical reaction. An ordinary catalyst may be used for several different chemical reactions, but an enzyme only works for one specific reaction.

Human saliva contains an enzyme called amylase. This enzyme helps to turn starch into a sugar called maltose. When you swallow a mouthful of food, the amylase stops working because it is much too acid in the stomach pH 2. Amyalse works best in neutral or slightly alkaline conditions, *i.e.*, at about pH 7.

When your food gets into the small intestine, more amylase is made by the pancreas and this turns the remaining starch into maltose. Another enzyme (maltase) turns all this maltose into glucose. Glucose is then absorbed into the blood.

Enzymes in the human alimentary canal and what they digest:

Enzyme	Substrate
Amylase	Starch
Maltase	Maltose
Sucrase	Sucrose
Lipase	Fats
Pepsin	Proteins

All animals, green plants, fungi and bacteria produce enzymes: so enzymes are not just about digesting food. The enzymes which we use to digest our food are extra-cellular, that means they are found outside cells. We also have enzymes inside our cells; these are intra-cellular enzymes. Enzymes are used in ALL chemical reactions in living things; this includes respiration, photosynthesis, movement growth, getting rid of toxic chemicals in the liver and so on.

Viruses are rather different, but you do not need to know much about them for GCSE, so just make sure that you don't catch any!

Enzymes must have the correct shape to do their job. They are made of proteins, and proteins are very easily affected by heat, pH and heavy metal ions. Some people say that enzymes work like a key in a lock.

If the key has been twisted by heat, or dissolved in acid or stuck up with chewing gum it will not work.

Enzymes change their shape if the temperature or pH changes, so they have to have the right conditions. Copper ions are poisonous: if you get copper ions in your blood they will block up some of the important enzymes in red and white blood cells.

Enzymes are catalysts. Most are proteins. (A few ribonucleoprotein enzymes have been discovered and, for some of these, the catalytic activity is in the RNA part rather than the protein part. Link to discussion of theseribozymes.)

Enzymes bind temporarily to one or more of the reactants of the reaction they catalyse. In doing so, they lower the amount of activation energy needed and thus speed up the reaction.

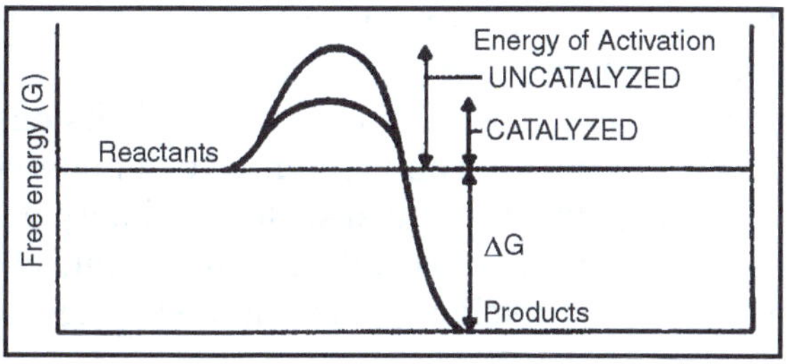

Examples:

- *Catalase*: It catalyses the decomposition of hydrogen peroxide into water and oxygen.

$$2H_2O_2 \rightarrow 2H_2O + O_2$$

One molecule of catalase can break 40 million molecules of hydrogen peroxide each second.

- *Carbonic anhydrase*: It is found in red blood cells where it catalyses the reaction

$$CO_2 + H_2O \leftrightarrow H_2CO_3$$

It enables red blood cells to transport carbon dioxide from the tissues to the lungs. One molecule of carbonic anhydrase can process one million molecules of CO_2 each second.

- *Acetylcholinesterase:* It catalyses the breakdown of the neurotransmitter acetylcholine at several types of synapses as well as at the neuromuscular junction — the specialised synapse that triggers the contraction of skeletal muscle.

One molecule of acetylcholinesterase breaks down 25,000 molecules of acetylcholine each second. This speed makes possible the rapid "resetting" of the synapse for transmission of another nerve impulse. In order to do its work, an enzyme must unite — even if ever so briefly — with at least one of the reactants. In most cases, the forces that hold the enzyme and its substrate are non-covalent, an assortment of:

- hydrogen bonds
- ionic interactions
- and hydrophobic interactions

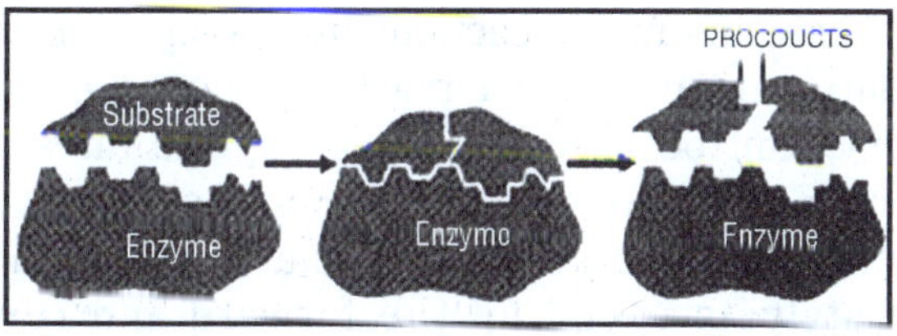

Most of these interactions are weak and especially so if the atoms involved are farther than about one angstromfrom each other. So successful binding of enzyme and substrate requires that the two molecules be able to approach each other closely over a fairly broad surface.

Thus the analogy that a substrate molecule binds its enzyme like a key in a lock. This requirement for complementarity in the configuration of substrate and enzyme explains the remarkable specificity of most enzymes. Generally, a given enzyme is able to catalyse only a single chemical reaction or, at most, a few reactions involving substrates sharing the same general structure.

COMPETITIVE INHIBITION

The necessity for a close, if brief, fit between enzyme and substrate explains the phenomenon of competitive inhibition.

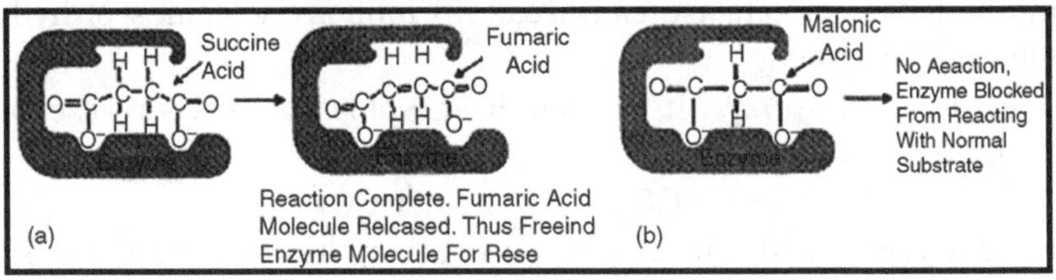

One of the enzymes needed for the release of energy within the cell is succinic dehydrogenase.

It catalyses the oxidation (by the removal of two hydrogen atoms) of succinic acid

(a) If one adds malonic acid to cells, or to a test tube mixture of succinic acid and the enzyme, the action of the enzyme is strongly inhibited. This is because the structure of malonic acid allows it to bind to the same site on the enzyme

(b) But there is no oxidation so no speedy release of products. The inhibition is called competitive because if you increase the ratio of succinic to malonic acid in the mixture, you will gradually restore the rate of catalysis. At a 50:1 ratio, the two molecules compete on roughly equal terms for the binding (=catalytic) site on the enzyme.

CELLS ORGANISMS OF ENZYMES AS BIOLOGICAL CATALYSTS

In cells and organisms most reactions are catalyzed by enzymes, which are regenerated during the course of a reaction. These biological catalysts are physiologically important because they speed up the rates of reactions that would otherwise be too slow to support life. Enzymes increase reaction rates, sometimes by as much as one million-fold, but more typically by about one thousand fold. Catalysts speed up the forward and reverse reactions proportionately so that, although the magnitude of the rate constants of the forward and reverse reactions is are increased, the ratio of the rate constants remains the same in the presence or absence of enzyme. Since the equilibrium constant is equal to a ratio of rate constants, it is apparent that enzymes and other catalysts have no effect on the equilibrium constant of the reactions they catalyze.

Enzymes increase reaction rates by decreasing the amount of energy required to form a complex of reactants that is competent to produce reaction products. This complex is known as the activated state or transition state complex for the reaction. Enzymes and other catalysts accelerate reactions by lowering the energy of the transition state. The free energy required to form an activated complex is much lower in the catalyzed reaction. The amount of energy required to achieve the transition state is lowered; consequently, at any instant a greater proportion of the molecules in the population can achieve the transition state. The result is that the reaction rate is increased.

Michaelis-Menten Kinetics

In typical enzyme-catalyzed reactions, reactant and product concentrations are usually hundreds or thousands of times greater than the enzyme concentration. Consequently, each enzyme molecule catalyzes the conversion to product of many reactant molecules. In biochemical reactions, reactants are commonly known as substrates.

The catalytic event that converts substrate to product involves the formation of a transition state, and it occurs most easily at a specific binding site on the enzyme. This site, called the catalytic site of the enzyme, has been evolutionarily structured to provide specific, high-affinity binding of substrate(s) and to provide an environment that favours the catalytic events. The complex that forms, when substrate(s) and enzyme combine, is called the enzyme substrate (ES) complex. Reaction products arise when the ES complex breaks down releasing free enzyme.

Between the binding of substrate to enzyme, and the reappearance of free enzyme and product, a series of complex events must take place. At a minimum an ES complex must be formed; this complex must pass to the transition state (ES*); and the transition state complex must advance to an enzyme product complex (EP). The latter is finally competent to dissociate to product and free enzyme.

The series of events can be shown thus:

$$E + S \leftrightarrow ES \leftrightarrow ES^* \leftrightarrow EP \leftrightarrow E + P$$

The kinetics of simple reactions like that above were first characterized by biochemists Michaelis and Menten. The concepts underlying their analysis of enzyme kinetics continue to provide the cornerstone for understanding metabolism today, and for the development and clinical use of drugs aimed at selectively altering rate constants and interfering with the progress of disease states. The Michaelis-Menten equation is a quantitative description of the relationship among the rate of an enzyme- catalyzed reaction [v_1], the concentration of substrate [S] and two constants, V_{max} and K_m (which are set by the particular equation). The symbols used in the Michaelis-Menten equation refer to the reaction rate [v_1], maximum reaction rate (V_{max}), substrate concentration [S] and the Michaelis-Menten constant (K_m).

$$V_1 = \frac{V_{max}[s]}{\{K_m + [S]\}}$$

The Michaelis-Menten equation can be used to demonstrate that at the substrate concentration that produces exactly half of the maximum reaction rate, *i.e.* ½ V_{max}, the substrate concentration is numerically equal to K_m. This fact provides a simple yet powerful bioanalytical tool that has been used to characterize both normal and altered enzymes, such as those that produce the symptoms of genetic diseases. Rearranging the Michaelis-Menten equation leads to:

$$K_m = [S]\left\{\left[\frac{V_{max}}{V_1}\right] - 1\right\}$$

From this equation it should be apparent that when the substrate concentration is half that required to support the maximum rate of reaction, the observed rate, v_1, will, be equal to V_{max} divided by 2; in other words, v_1 = [V_{max}/2]. At this substrate concentration V_{max}/v_1 will be exactly equal to 2, with the result that:

$$[S](1) = K_m$$

The latter is an algebraic statement of the fact that, for enzymes of the Michaelis-Menten type, when the observed reaction rate is half of the maximum possible reaction rate, the substrate concentration is numerically equal to the Michaelis-Menten constant. In this derivation, the units of K_m are those used to specify the concentration of S, usually Molarity.

The Michaelis-Menten equation has the same form as the equation for a rectangular hyperbola; graphical analysis of reaction rate (v) versus substrate concentration [S] produces a hyperbolic rate plot.

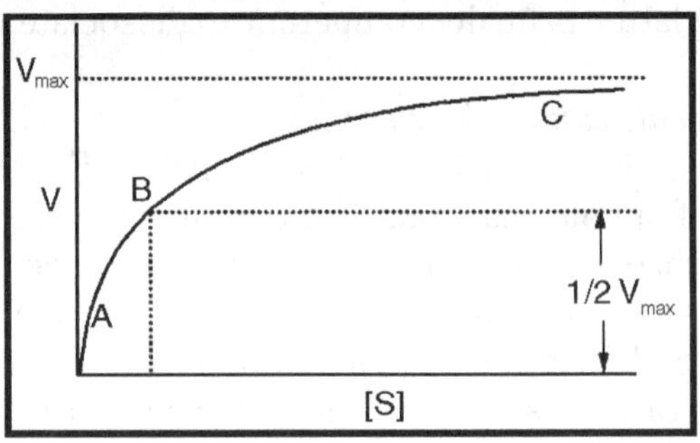

Fig. Plot of Substrate Concentration Versus Reaction Velocity

The key features of the plot are marked by points A, B and C. At high substrate concentrations the rate represented by point C the rate of the reaction is almost equal to V_{max}, and the difference in rate at nearby concentrations of substrate is almost negligible.

If the Michaelis-Menten plot is extrapolated to infinitely high substrate concentrations, the extrapolated rate is equal to V_{max}. When the reaction rate becomes independent of substrate concentration, or nearly so, the rate is said to be zero order. (Note that the reaction is zero order only with respect to this substrate. If the reaction has two substrates, it may or may not be zero order with respect to the second substrate).

The very small differences in reaction velocity at substrate concentrations around point C (near V_{max}) reflect the fact that at these concentrations almost all of the enzyme molecules are bound to substrate and the rate is virtually independent of substrate, hence zero order. At lower substrate concentrations, such as at points A and B, the lower reaction velocities indicate that at any moment only a portion of the enzyme molecules are bound to the substrate. In fact, at the substrate concentration denoted by point B, exactly half the enzyme molecules are in an ES complex at any instant and the rate is exactly one half of V_{max}. At substrate concentrations near point A the rate appears to be directly proportional to substrate concentration, and the reaction rate is said to be first order.

Inhibition of Enzyme Catalyzed Reactions

To avoid dealing with curvilinear plots of enzyme catalyzed reactions, biochemists Lineweaver and Burk introduced an analysis of enzyme kinetics based on the following rearrangement of the Michaelis-Menten equation:

$$\frac{1}{V} = \left[\frac{K_m(1)}{V_{max}(S)} + \frac{1}{V_{max}}\right]$$

Plots of 1/v versus 1/[S] yield straight lines having a slope of K_m/V_{max} and an intercept on the ordinate at $1/V_{max}$.

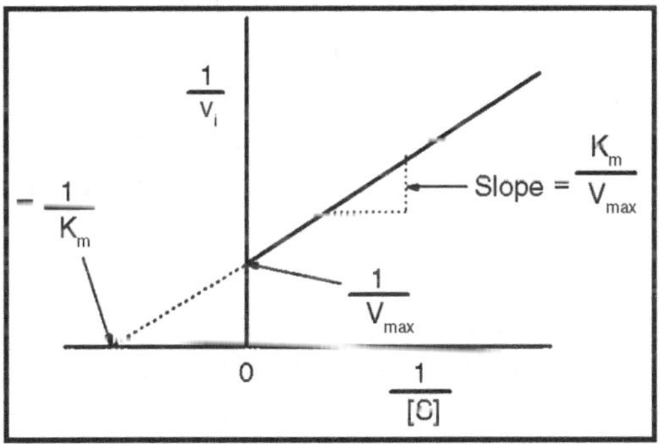

Fig. A Lineweaver-Burk Plot

An alternative linear transformation of the Michaelis-Menten equation is the Eadie-Hofstee transformation:

$$\frac{v}{[S]} = -v\left[\frac{1}{K_m} + \right] + \left[\frac{V_{max}}{K_m}\right]$$

and when v/[S] is plotted on the y-axis versus v on the x-axis, the result is a linear plot with a slope of $-1/K_m$ and the value V_{max}/K_m as the intercept on the y-axis and V_{max} as the intercept on the x-axis.

Both the Lineweaver-Burk and Eadie-Hofstee transformation of the Michaelis-Menton equation are useful in the analysis of enzyme inhibition. Well-known examples of such therapy include the use of methotrexate in cancer chemotherapy to semi-selectively inhibit DNA synthesis of malignant cells, the use of aspirin to inhibit the synthesis of prostaglandins which are at least partly responsible for the aches and pains of arthritis, and the use of sulfa drugs to inhibit the folic acid synthesis that is essential for the metabolism and growth of disease-causing bacteria. In addition, many poisons, such as cyanide, carbon monoxide and polychlorinated biphenols (PCBs) produce their life-threatening effects by means of enzyme inhibition.

Enzyme inhibitors fall into two broad classes: those causing irreversible inactivation of enzymes and those whose inhibitory effects can be reversed. Inhibitors of the first class usually cause an inactivating, covalent modification of enzyme structure.

Cyanide is a classic example of an irreversible enzyme inhibitor: by covalently binding mitochondrial cytochrome oxidase, it inhibits all the reactions associated with electron transport.

The kinetic effect of irreversible inhibitors is to decrease the concentration of active enzyme, thus decreasing the maximum possible concentration of ES complex.

Since the limiting enzyme reaction rate is often k_2[ES], it is clear that under these circumstances the reduction of enzyme concentration will lead to decreased reaction rates. Note that when enzymes in cells are only partially inhibited by irreversible inhibitors, the remaining unmodified enzyme molecules are not distinguishable from those in untreated cells; in particular, they have the same turnover number and the same K_m.

Turnover number, related to V_{max}, is defined as the maximum number of moles of substrate that can be converted to product per mole of catalytic site per second.

Irreversible inhibitors are usually considered to be poisons and are generally unsuitable for therapeutic purposes.

Reversible inhibitors can be divided into two main categories; competitive inhibitors and non-competitive inhibitors, with a third category, uncompetitive inhibitors, rarely encountered.

Inhibitor Type	Binding Site on Enzyme	Kinetic effect
Competitive Inhibitor	Specifically at the catalytic site, where it competes with substrate for binding in a dynamic equilibrium-like process. Inhibition is reversible by substrate.	V_{max} is unchanged; K_m, as defined by [S] required for ½ maximal activity, is increased.
Non-competitive Inhibitor	Binds E or ES complex other than at the catalytic site. Substrate binding unaltered, but ESI complex cannot form products. Inhibition cannot be reversed by substrate. concentration.	K_m appears unaltered; V_{max} is decreased proportionately to inhibitor
Uncompetitive Inhibitor	Binds only to ES complexes at locations other than the catalytic site. Substrate binding modifies enzyme structure, making inhibitor- binding site available. Inhibition cannot be reversed by substrate.	Apparent V_{max} decreased; K_m, as defined by [S] required for ½ maximal activity, is decreased.

The hallmark of all the reversible inhibitors is that when the inhibitor concentration drops, enzyme activity is regenerated. Usually these inhibitors bind to enzymes by non-covalent forces and the inhibitor maintains a reversible equilibrium with the enzyme. The equilibrium constant for the dissociation of enzyme inhibitor complexes is known as K_i:

$$K_i = \frac{[E][I]}{[EI]}$$

The importance of K_I is that in all enzyme reactions where substrate, inhibitor and enzyme interact, the normal K_m and or V_{max} for substrate enzyme interaction appear to be altered. These changes are a consequence of the influence of K_i on the overall rate equation for the reaction. The effects of K_i are best observed in Lineweaver-Burk plots.

Probably the best known reversible inhibitors are competitive inhibitors, which always bind at the catalytic or active site of the enzyme. Most drugs that alter enzyme activity are of this type. Competitive inhibitors are especially attractive as clinical modulators of enzyme activity because they offer two routes for the reversal of enzyme inhibition, while other reversible inhibitors offer only one. First, as with all kinds of reversible inhibitors, a decreasing concentration of the inhibitor reverses the equilibrium regenerating active free enzyme. Second, since substrate and competitive inhibitors both bind at the same site they compete with one another for binding.

Raising the concentration of substrate (S), while holding the concentration of inhibitor constant, provides the second route for reversal of competitive inhibition. The greater the proportion of substrate, the greater the proportion of enzyme present in competent ES complexes. As noted earlier, high concentrations of substrate can displace virtually all competitive inhibitor bound to active sites. Thus, it is apparent that V_{max} should be unchanged by competitive inhibitors. This characteristic of competitive inhibitors is reflected in the identical vertical-axis intercepts of Lineweaver-Burk plots, with and without inhibitor.

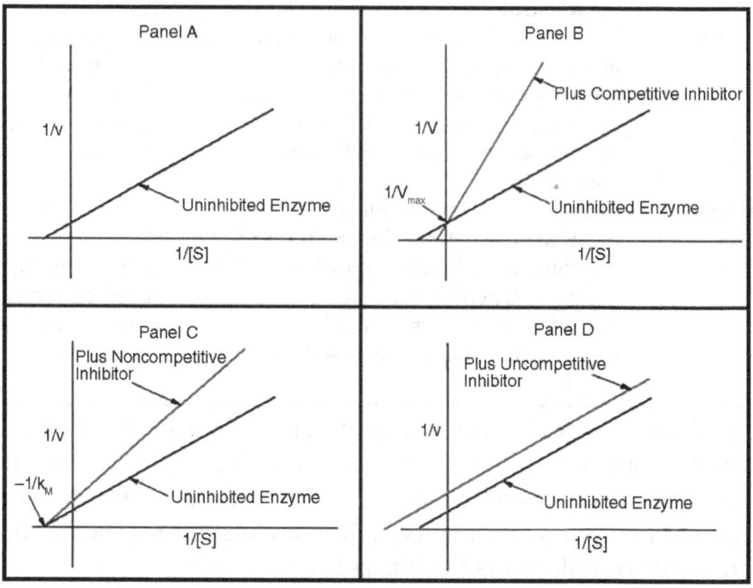

Fig. Lineweaver-Burk Plots of Inhibited Enzymes

Since attaining V_{max} requires appreciably higher substrate concentrations in the presence of competitive inhibitor, K_m (the substrate concentration at half maximal velocity) is also higher, as demonstrated by the differing negative intercepts on the horizontal axis in panel B.

Analogously, panel C illustrates that non-competitive inhibitors appear to have no effect on the intercept at the x-axis implying that non-competitive inhibitors have no effect on the K_m of the enzymes they inhibit. Since non-competitive inhibitors do not interfere in the equilibration of enzyme, substrate and ES complexes, the K_m's of Michaelis-Menten type enzymes are not expected to be affected by non-competitive inhibitors, as demonstrated by x-axis intercepts in panel C. However, because complexes that contain inhibitor (ESI) are incapable of progressing to reaction products, the effect of a non-competitive inhibitor is to reduce the concentration of ES complexes that can advance to product. Since $V_{max} = k_2[E_{total}]$, and the concentration of competent E_{total} is diminished by the amount of ESI formed, non-competitive inhibitors are expected to decrease V_{max}, as illustrated by the y-axis intercepts in panel C.

A corresponding analysis of uncompetitive inhibition leads to the expectation that these inhibitors should change the apparent values of K_m as well as V_{max}. Changing both constants leads to double reciprocal plots, in which intercepts on the x and y axes are proportionately changed; this leads to the production of parallel lines in inhibited and uninhibited reactions.

CATALYSES STRUCTURE OF PHOSPHOFRUCTOKINASE IN ENZYMES

Structure of phosphofructokinase (wire-frame), with its substrates and the allosteric effector AMP bound to it (space-fill). In Figure the substrates are located next to each other within the active site of the dimeric enzyme. This reaction is an early step in the degradation of glucose, which ultimately serves to replenish ATP from ADP and phosphate.

It therefore makes good biological sense that phosphofructokinase should be stimulated by ADP. To accomplish this stimulation, ADP binds to another site that is far away from the active site; it therefore clearly does not directly participate in the reaction. Instead, ADP binding changes the conformation of the entire enzyme molecule. The change will also affect the active site and enhance the efficiency of catalysis there. This mode of action is known as allosteric regulation and is exceedingly common. Allostericeffectors can be either stimulatory (as ADP is in this example) or inhibitory. *E.g.*, ATP does not only bind to the active site but also acts as an allosteric inhibitor of phosphofructokinase.

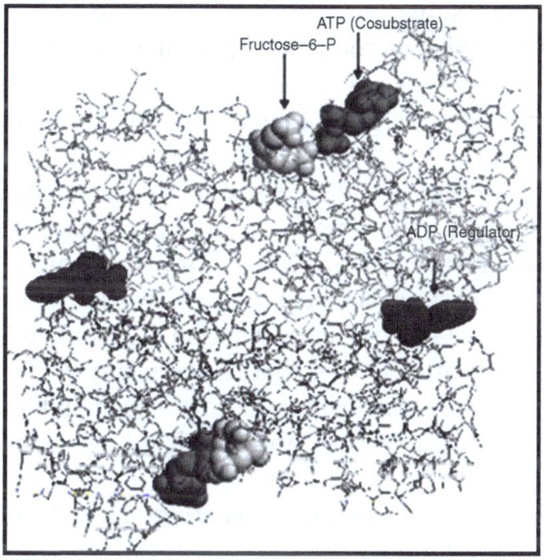

Fructose-6-phosphate + ATP '!
Fructose -1,6-bisphosphate + ADP

The workings of allosteric regulation are schematically depicted Figure The enzyme has two possible conformations that are in equilibrium with each other An allosteric activator will bind selectively to the regulatory site as it occurs in the active conformation and thereby shift the equilibrium towards this conformation. Conversely, an inhibitor would bind selectively to the inactive conformation and thereby stabilise it. As you can see, activators and inhibitors may share the same regulatory site; this is the case in the above example of phosphofructokinase with ATP and ADP. Note, however, that phosphofrucokinase has additional allosteric sites that permit regulation by other effectors.

Enzyme regulation by allosteric effectors and by phosphorylation. a: The binding site for the allosteric effector (RS) is distinct from the active site (AS) of the enzyme.

The enzyme can assume two distinct conformations, and both the regulatory site and the active site are affected by the transition between them. Only in one conformation the enzyme is active. b: An inhibitor binds to a regulatory site in the inactive conformation, and it will change the conformational equilibrium accordingly. An activator stabilises the active conformation. c: Most allosteric enzymes are multimeric, and their transitions are synchronous. d: Regulation by phosphorylation also works by selective stabilisation of one conformation. Like allosteric regulation, it can be inhibitory (as shown here) or stimulatory.

Although it is not theoretically necessary, it seems that all allosteric enzymes occur as oligomers. In Figure dimeric enzyme is shown, but often the number of subunits is considerably higher. Their oligomeric nature enables enzymes to react more sensitively to changes in effector concentration.

Regulation of phosphofructokinase and of fructose-1,6-bisphosphatase. a: Allosteric effectors have opposite effects on the two enzymes. b: Structure of fructose-2,6-bisphosphate (right). Fructose-1,6-bisphosphate is shown for comparison. c: Control of fructose-2,6-bisphosphate levels by hormones.

Another important means of enzyme regulation consists in *phosphorylation*. This occurs by protein kinases, which transfer a phosphate group from ATP to a specific site on the regulated enzyme. The mechanism of regulation by phosphorylation is not really that different – it also works by selective stabilisation of either the active or the inactive conformation. The only difference is that the regulator is more stably attached, so that the regulatory effect may last longer. Like allosteric regulation, it can be inhibitory (as shown in Figure or stimulatory. Many enzymes are subject to regulation both by allosteric effectors and by phosporylation.

While all mechanisms discussed so far modulate the activity of existing enzyme molecules, the overall enzyme activity may also be varied by changing their abundance:

- The transcription of the gene encoding the enzyme in question can be turned on or off. This mechanism is employed by many hormones, in particular steroid hormones (*e.g.*, cortisone) and thyroid hormones.
- The stability of the enzyme mRNA's of the enzymes can be changed.

Hormones may affect the activity of an enzyme at more than one level. For example, insulin increases the activity of glycogen synthase by way of transcriptional induction, increased mRNA stability, and protein phosphorylation.

ENZYMES

In such manner Kossel and Dakin * freed liver arginase from most proteins. Essentially the same principle applies when fresh tissue-extracts are al- lowed to stand for some time; coagulation being then due to he weak acid reaction resulting in muscles and other tissues shortly after death.

If we dialyse the extract at this time in running water, we remove both the proteins which become insoluble and the salts and other crystalloid substances which pass through the parchment. After the dialysis, or after standing, we filter oflE the precipitate.

As a rule it is useless to filter under pressure through a Chamberland or similar filter. Nevertheless, the advantage of Buchner's method con- sists not only in the complete disintegration of tissues and structures, but likewise in the filtration of the press- juice through siliceous earth, which retains the nucleoproteins, the globulins, and all proteins coagulating spontaneously. Almost the only proteins found in the press-juice of the muscles and the liver are albumin and the haemoglobin of the blood. These juices are in most cases very rich in enzymes.

Using one of these methods it is easy to obtain proteolytic enzymes and steapsins, but many sucroclastic enzymes are retained by the proteins, and cannot be dissolved out again by water. Sometimes it is possible to dissolve them by very weak alkali, but the necessary conditions have not been suflSciently studied.

If we do not succeed in bringing the enzymes into solution, we must add the solid precipitate to the solution containing carbohydrates or fats or proteins, which are to be acted upon by the enzyme. We do not know whether the enzyme works under such conditions in a solid state and converts carbohydrates and other substances only by contact, or whether the enzyme passes into solution.

We shall see that we must assume a chemical connection and combination between enzyme and substance. Per- haps the chemical compound,thus newly formed is more soluble than the free enzyme.

This would be a good ex- planation of the action, for instance, by which the mucous membrane of the small intestine converts milk sugar, although it is impossible to find lactase in the extracts, or the ability of dried pancreas powder to dissolve protein,which is lacking in extracts made from this powder. We must always remember, however, that the difference can be produced also by an improper reaction of the extract. It is an important fact, however, that we can use the dry powder for many purposes.

With these methods we can easily remove proteins and other harmless, indifferent substances. The chief diffi- culties which complicate the purification, extraction, and investigation of enzymes, are the presence of other enzymes, and the anchorage of the enzymes in the cells. The avoidance of these difficulties is the chief advantage of the secreted natural juices as compared with the extracts.

Another without great loss. The gastric juice contains only the well-known proteolytic enzyme, pepsin, which dissolves natural proteins and converts them into peptones; but the process never goes beyond the stage of yielding biuret-giving substances.

The mucous membrane of the stomach, and therefore its extracts, contains, besides pepsin destined for secretion, an erepsin acting intracellularly and splitting peptones into amino-acids. The pepsin and the erepsin have the same solubilities, and if we free the proteolytic enzyme from impurities, we concentrate both enzymes, getting an artificial juice with properties other than those of the natural one. The con- tradictions of authors on the question of trypsin and trypsinogen is believed by Bayliss * to be susceptible of the same explanation.

If we microscopically study the pancreas or a salivary gland at rest, we see the cell crowded with small, refractive granules forming a mass that leaves only a very narrow clear space next to the basement membrane. In the salivary gland the cell can be wholly studded with the granular mass. During glandular activity the granfiles become much fewer in number, and retract to an inner narrow margin; in the salivary gland they may even dis- appear. There can be no doubt that during the act of secretion, the granules are discharged to form part of the secretion.

They become the solid matter of the juices; therefore they must be soluble in water. If we extract the pancreas or the parotid gland while studded with these granules, they, and with them the enzymes, are dissolved immediately. As has been pointed out by Pekelharing, ^ we find in the extracts of the gastric mucous membrane the most characteristic chemical compound of the gastric juice.

There is something else to be noticed in dealing with the intracellular enzymes that never leave the cell. In the state of activity, both these enzymes and the substances hydrolysed or oxidised by them must be in solution, for "corpora non- agunt nisi soluta."

To avoid the action of enz)anes at rest, the two bodies must be separated from each other; they must be fixed outside the real fluid protoplasm. We know another substance which, in this respect, is analogous to the endo-enzymes; glycogen, which is soluble in water, but which may be stored in insoluble form within the cell. Ehrlich ^ found that glycogen is fixed in the liver cell by a "Trager- substanz," a holding-substance, and that one of the colour reactions attributed to glycogen is due to this substance. Glycogen and this Tragersubstanz are rather firmly bound together. We know with what difficulty the glycogen is completely dissolved out of the organs, and with what difficulty it is freed from the last traces of im- purities. It seems to me that we must also assume such a holding-substance for the intracellular enzymes which restrains the action of the enzyme even after the almost complete disintegration of the tissues.

CHEMICAL PROCESSES OF ENZYMES CATALYSERS

We know of chemical processes in whicli the reacting substances completely disappear, giving rise to a new body. But in other processes one chemical compound reacts with another compound and changes it without entering into the final reaction itself. For instance, the conversion of cane sugar is expressed by the equation:

$$C_{12}H_{22}O_{11} + H_2O + wHCl = CeHijOe + CeHizOe + wHCL$$

The hydrochloric acid causes the reaction, but before and after the reaction we find the same quantity of acid. A second case is the formation of ether from alcohol by sulphuric acid:

$$2C_2H_5OH + WSO_4H_2 = C_2H_5.. C_2H_5 + H_2O + WSO_4H_2.$$

The sulphuric acid is the cause of the reaction, but neither the original alcohol nor the end-products of the reaction contain the sulphuric acid, and the most import- ant point is the fact that there is no quantitative relation between the hydrochloric acid and the sugar or between the sulphuric acid and the alcohol and ether. Chemists * apply the term "catalytic" to such reactions, and it seems that all enzymotic processes are catalytic reactions.

The evidence for their designation as catalysers is that we cannot find any relation between the quantity of enzyme and the quantity of matter decomposed by it, and that, "> ^ after the reaction, we find the enzyme is not consumed. The last argument, however, is not completely proved, because the estimation of the enzymes is not so exact as to exclude the consumption of a portion.

The great discrepancy between the small quantity of enzymos and the enormous quantity of matter converted by enzymes, is astonishing, and has impressed ail in- vestigators. The best explanation seems to be the admission of an intercalated reaction assumed by chemists generally in the case of thc formation of ether.

Thus the reaction would be expressed by two equations:

- $C_2H_5OH + SO.H^\wedge = C_2H_5$ (HSOJ + H_2O).
- C_2H_5 (HSO J + $C_2H_5OH = C_2H_5.. C_3H, + II_3SO$.

That is to say, the molecule of thc sulphuric acid is combined first with one molecule of alcohol, then it combines with a second molecule of alcohol, and finally it is liberated and can attack a new molecule of alcohol. It is highly probable that enzymes are combined thus with the substrate they work upon and with the products they yield.

The arguments for this supposition will be cited later on, but since I have spoken of the enzymes as cataly- zers, the theory of Ostwald and Bredig should also be mentioned. They have ascribed a new function to cata- lysers besides the quality mentioned.

They define a catalyser as a substance that never causes a reaction, but only accelerates a reaction going on spontaneously but more slowly without the catalyser. They propose the theory that the proteins or carbohydrates are always slowly dissociated in watery solution, and that the enzymes hasten this dissociation.

They thought that through this theory they could avoid the difficulty presented by the fact that enzymes mediate in such extensive processes, extensive also as regards the energy involved. They argue that our body produces its whole energy through the enz)mies oxidising and burning the food, and that it is difficult to suppose that the great quantity of energy four large calories per granmie of glucose, and nine per granmie of fat - can be liberated by the small quantity of enzymes which are available.

The argument of Ostwald and Bredig has been adopted by many investigators; in fact we may say that it has ruled the whole theory of enzymotic action in recent years, but it seems to me that it does not hold good, and for three reasons:

- It is not true that proteins or carbohydrates spon- taneously undergo a slow dissociation in solution. Such solutions have been preserved for seventeen years or more at room temperature, and, in the absence of micro-organ- isms, of acid, or of alkali, the substances have been found unchanged. We have no reasons for thinking that the time of obser\"ation is too short, and that we would find a dissociation in centuries.
- If we heat cane sugar with acid, it is decomposed in the same manner as by invertin, and we could here attribute the decomposition really to the H-ions, which neutral water contains in lowest concentration, and assume simply that the enzyme hastens it. But in other cases, the enzymes cause other reactions than the H-ions of the water would bring about. The dissociation of glucose produced by H- or OH-ions gives rise to lactic acid and humins, while the dissociation by zymase produces alcohol.
- Sugars and fats are oxidised and burned in the tissue, and set free much energy, but we do not know exactly to what extent this oxidation is caused by enzymes, and especially by one enzyme. It is possible that the structure of cells is needed for the complete combustion, and that two, three, or more enz3mies work

consecutively and perhaps in separate localities. We know that Z3miase is the only enzyme that gives off heat, and only a very small quantity of heat. The well-known enzyiaes, the hydrolytic ones, liberate no heat. The action of trypsin, was measured by Grafe ^ in Rubner's laboratory with a calorimeter of high sensitiveness. He found no rise of temperature.

- The other enzymes have not been measured in this manner, so far as I know, but the heat of combustion of starch, the disaccharides, and of glucose, has been determined, and we can calculate that the differences lie within the errors of analysis. The building-up of protein from the amino-acids, or of the polysaccharides from the hexoses, as well as the accompan)dng dissociations, seems to be unconnected with any perceptible thermic process. It is of great importance for the understanding of enzymes, that the hydrolytic enzymes neither give out nor consume any considerable or perceptible quantities of heat. Opin- ions regarding energy have here no weight. We must see whether well-observed facts support the theory that enzymes are like inorganic catalysers, and only hasten reactions without provoking them. For all proteolytic enzymes no fact is known which gives any evidence for it, and the same is true of ptyalin, invertin, lactase, zymase, and the corresponding metabolism enzymes urease and nuclease.
- On the other hand, evidence has been brought forward that esters like fats are dissociat- ed spontaneously by the ions of the water, and that steap- sins or lipases which split up fats and other esters, only hasten this process. Later on, in treating of the synthetical action of enzymes, I shall discuss the steapsins, or lipases, which are distinguished from all other enzymes, because they alone effect synthesis to any great extent.
- I think that the capacity for acting synthetically is connected with the property of the steapsins to hasten only a slow spon- taneous action. The steapsin seems to have the proper- ties of catalysers in the sense of Ostwald and Bredig, and this property lies at the bottom of the synthetical action. According to the most important work of Pottevin, ^ the relations between dissociations and synthetical action in steapsin are perfectly similar to the relations between alcohols, acids, and esters in watery solution without enzymes, because both reactions show an equilibrium- point, which is moved in the one or the other direction according to the temperature or the presence of more or less water. If ethyl alcohol is heated with acetic acid, the following reaction occurs:

$$CH_3COOH + C_2H_5OH = CH_3CO.O.C_3H_5 + HA$$

But if water and ethyl acetate are mixed, the opposite reaction takes place, as follows:

$$CH_3CO.O.C_2H_5 + H,0 = CH_3COOH + C_2H_5OH.$$

If these two substances, alcohol and acetic acid, are allowed to mix, both reactions must proceed simultaneously in the same solution until they reach an equilibrium-point, that is to say, the point at which both proceed at the same rate, and all change seems to be stopped. The position of the equilibrium-point depends upon the mass of the three reacting substances, water, acid, and alcohol, and it depends, further, upon the temperature of the solution. The equation is therefore expressed in this manner:

$$CH_3COOH + C_2H50H?=\wedge CH_3CO.O.C_3H_5 + HA$$

The steapsin moves the equilibrium-point, and thus acts like a true catalyser. In concentrated solutions it hastens s)mthesis, but if we add water, it hastens disso- ciation. It is remarkable now, that the disaccharides, maltose and isomaltose, are glucosides according to E. Fischer, and are similar to ether in structure, and that he even has observed strong hydrochloric acid to have a synthetical influence on glucose.

Compounds like isomaltose are thus formed. No evidence has been brought forward thus far regarding the slow dissociation of maltose in the absence of maltase, but perhaps we can expect this fact to be observed, and in that case maltase must be likened to steapsin and not to sucroclastic enzymes which attack the anhydrous polysaccharides like starch or cane sugar, and which exhibit no catalytic quality.

A second class of enzymes, which hasten only those reactions proceeding slowly without them, are the oxidase;: of the type of laccase. The more important metabolic enzymes, like zymase or lactacidase, I repeat emphatically, provoke conversions which do not occur at all in the absence of enzymes, and so do all proteolytic enzymes and the nucleases. Bredig, Henri, and others have compared the time velocity of reactions, which are produced by inorganic catalysers and by enzymes, and they have found that the curves resemble one another, but show some dififerences.

I have already stated that it is impossible to study the laws of action, if we have only solutions so crowded with impurities as are our enzyme solutions. I have said, and may repeat again, that tnzyiaes are modified in all sorts of ways in solutions, and shall presently mention that the action of enzymes is checked by the products formed by their action, and occurring therefore in the solutions; while in the case of ether or cane sugar, the adds pro- voking reactions are not influenced by. substances appeaing in the solution. Thus irregular curves result, which throw light upon the happenings in enzymotic processes.

We can only observe that the action of enz5anes begins immediately when they touch their substrate, and that the process, with suitable reaction and temperature, proceeds most rapidly. If we add saliva, containing ptyalin, to starch, reducing sugars can be detected as soon as the fluids are mixed. Fibrin is dissolved in gastric juice, containing pepsin, as rapidly as sugar dissolves in water. If we allow active trypsin to act upon casein or other easily digestible protein, tyrosine is split off in so short a time that it resembles a mere precipitation or crystallisation of the insoluble amino-acid. When study- ing arginase, Kossel wished to heat a liver extract with an enzyme in order to destroy arginase, and he set the tube in boiling water. The short time before the extract became heated was sufficient for the arginase to act upon the arginine and to split off some urea.

The glycolytic enzyme of muscles does not convert more sugar in twelve hours than in two. The quick beginning and the rapid progress is a special feature of enzymotic action. If I find a ferment that acts only after a long time, proceeds slowly, and forms end-products only in small quantity, I always conclude that either the enzyme is of no great importance, or is not present under suitable conditions.

TRIPLE ACTION CATALYSTS

Trifunctional organocatalysts that closely mimic natural enzymes can significantly increase reaction rates, say chemists in Japan. Tadashi Ema, Takashi Sakai and co-workers at University in Japan have made highly active organocatalysts by combining the reactivity of three separate organic components.

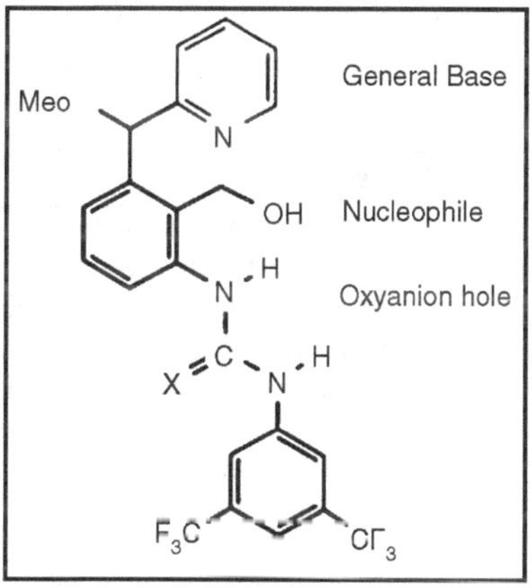

Fig. Catalytic Triad Structure of Many Enzymes Used as an Inspiration for an Organocatalyst

Mimicking the catalytic triad structure found inside many enzyme active sites, the three components work cooperatively, accelerating a test reaction 3.7 million-fold. Ema's catalysts were inspired by lipase, an enzyme known to work through a catalytic triad. 'With precise information about the structures and reaction mechanism of lipases, we decided to test our abilities to mimic the active site,' said Ema. The resulting trifunctional organocatalyst has a nucleophilic OH group, a pyridine moiety, and either a urea or thiourea group. Ema tested his catalysts using the transesterification reaction, a synthetic transformation often applied to organic molecules to switch one ester side chain for another.

The catalyst's hydroxyl group, boosted in reactivity by the adjacent pyridine, initiates the reaction by attacking the ester group in vinyl trifluoroacetate - and the urea group stabilises the resulting intermediate through hydrogen bonding.

Adding the catalyst increased the rate of reaction up to 3.7 million times, depending on the exact substrate used. For comparison, Ema also made three bifunctional control catalysts each missing a different one of the three components - and found all three to be essentially inactive. 'The observed rate accelerations are impressive,' said Ben List, who develops organocatalysts at the Institute. '

The authors have elegantly combined previously developed organocatalysis motifs to make a highly active and promising new organocatalyst type. Designing an asymmetric version is clearly the next logical step, which I look forward to.' Ema agrees that, along with further tuning the catalyst structure to boost reactivity levels even closer to those of enzymes, the next target will be to develop the catalyst for asymmetric synthesis.

ENZYMES AND INORGANIC CATALYSTS

Enzymes resemble inorganic catalysts in several aspects and differ from them in many other features. These similarities and differences are summarised in the following table.

Similarities	Diffferences
• Both are needed in imnute quantities	• Enzymes are complex proteins while catalysts are simple inorganic molecules
• Both accelerate the	• Enzymes have a very high rate of a rection, but molecular weight while cannot initiate one catalysts have a low molecular weight
• Both of them bring about a decrease in the activation energy	• Enzymes catalyse only biological reactions
• Bothe of them tempoarily combine with the substrate molecule	• Enzymes cataljyse only biological reactions
• Both of them do not undergo any chang in their composition	• Enzymes catalyse specific types kof reactions while catalysts have a wide range and hence, can be used again and agin
• Both of them do not	• Enzymes are regulated by

alter the nature and quantity of the end products	specific substances called cofactors, but not so in the case of catalysts
• The reaction accelerated by both of them is reversible	• Enzymes can be easily inactivated but not so in the case of catalysts

Mode of Enzyme Action

There are two different views to explain the mode of enzyme action the lock and key hypothesis and the induced-fit hypothesis.

Lock and Key Hypothesis

It was suggested by Emil Fischer in 1894. According to this view, the enzyme molecule operates by chemically uniting with the substrate molecule, forming an enzyme-substrate complex. The enzyme molecule provides a uniquely structured template on which the substrate molecules can become attached and interact subsequently.

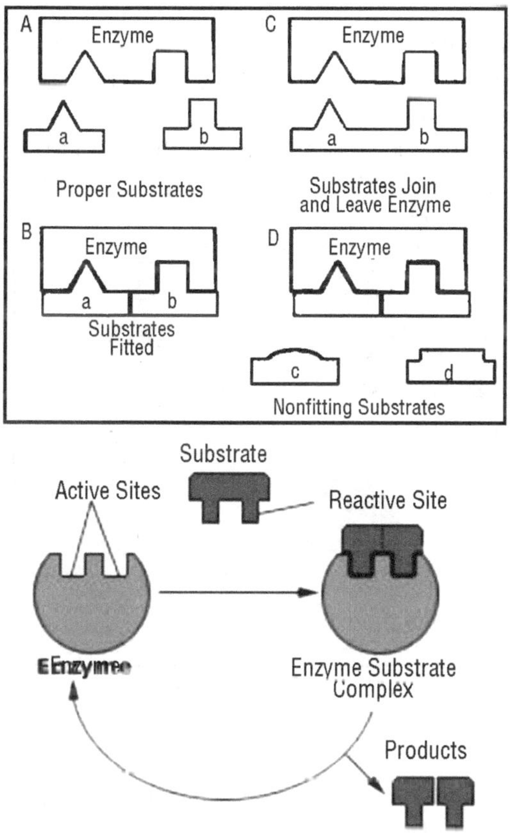

Fig. Breakdown Reaction Catalysed by an Enzyme According to Lock -and Key Hypothesis

This brings about an interaction between the specific active sites in the enzyme molecule and the reactive sites in the substrate molecule. The enzyme now breaks down the substrate into products. The products initially remain attached to the enzyme for a short while forming an enzyme product complex. The products get released from the enzyme molecule subsequently. The enzyme is now ready to receive another substrate molecule again. Thus, the same enzyme can be used again and again.

S	+	E	$\rightleftharpoons$	ES	$\rightleftharpoons$	EP	$\rightleftharpoons$	E + P
Substrate		Enzyme		Enzyme Substrate Complex		Enzyme Product Complex		Enzyme Product

Fig. - Lock and Key Hypothesis

Induced Fit Hypothesis

It was proposed by Daniel Koshland in 1959. According to this hypothesis, there is an intermediate condition called transition state between the substrate and the products.

It is highly unstable. When the substrate molecules bind to the enzyme molecule, a change is brought about in the active site to precisely fit the transition state (induced fit). This induced fit hold the substrates at the correct angle for the reaction to take place. The fact that an active site may also have a conformation to fit the product helps in explaining the role of enzymes in catalysing reversible reactions.

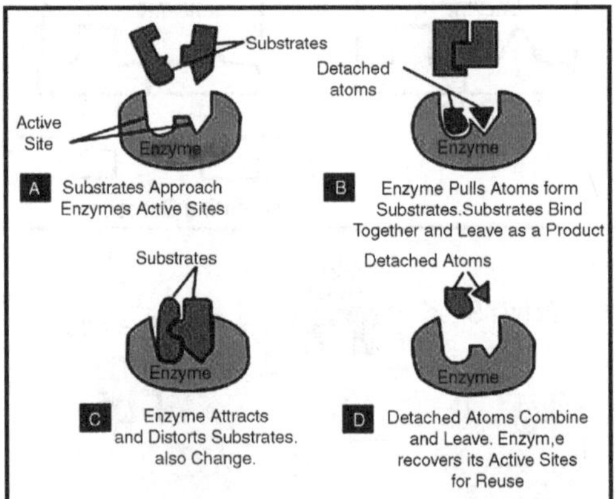

Working of an Enzyme According to the Induced Fit Hyptothesis

Fig. - Induced Fit Hypothesis

Isoenzymes

There are some enzymes which have slightly different molecular structure but exert similar catalytic action. Such enzymes are called isoenzymes or isozymes. More than 100 isozymes have been identified. The enzyme lactic dehydrogenase (LDH) in human skeletal muscle has five isozymes.

Subunit Association

Many enzymes can exhibit their catalytic activity only when they are in association with some non-protein substance. Such substances are called cofactors. The cofactors may be simple metal ions like Mg^{++} or complete organic compounds. An enzyme which functions only in the presence of a cofactor, is called apoenzyme. If the cofactor is an organic compound which can be easily separated from the apoenzyme, it is called coenzyme. If the cofactor is firmly bound with the enzyme, it is called prosthetic group. A working combination of an apoenzyme and its cofactor is called as holoenzyme.

BIOLOGICALLY FUNCTIONAL SURFACES OF ENZYMES

Biomaterials, especially when used in tissue engineering applications, must have the capacity to induce tissue regeneration/repair in order to achieve a more rapid recovery of the defect. At present, the existing scaffolds are not satisfactory in achieving rapid and full recovery of the defect. The attachment of cells to biomaterials, and their subsequent spreading, are mediated by extracellular matrix (ECM) glycoproteins such as fibronectin, collagen, etc. ECM glycoproteins contain short sequences with cell attachment properties, which interact with the integrin family of cell surface receptors. The peptide sequence Arg-Gly-Asp (RGD) has been identified as being capable of interacting with cell surface receptors.

Thus, the immobilization of biologically active molecules on the surface of biomaterials for presenting effectors to target cells or to induce a particular effect is of great interest, since the immobilization of active agents presents the advantage of providing a continuous and localized stimulus for cell proliferation. Unlike non-immobilized active agents, which are often consumed by cells, immobilized biomolecules remain bound to a substrate that is not consumed by cells and thus remain available to stimulate growth of additional cells.

This is particularly useful in perfusion cultures in which growth medium is continuously added and removed to allow long-term cell proliferation. Biodegradable polymers have been used as scaffolding materials for various tissue engineering applications because they can provide the support on which cells and tissues can adhere, but they can also guide and regulate the proliferation and activities of the adhered cells. However, the intrinsic hydrophobic property of some of these polymers restricts their applications as cell colonizing materials. Many methods have been used to modify the properties of polymer surface, such as plasma treatment-induced grafting polymerization, ozone oxidation, and immobilization of enzymes, and special biologically active agents have been used to introduce reactive groups onto polymeric surfaces. Various strategies have been developed to incorporate bioactive agents on the surface of biomaterials for controlling cell and tissue responses. The immobilization process can be involved, enriching surfaces for enhancing the cellular adhesion. Biomolecules such as enzymes, antibodies, antigens, peptides, or drugs have been immobilized on or within polymeric systems. An example of an adhesive protein is fibronectin, which is able to promote cellular adhesion through binding to integrin receptors, and this interaction has also been shown to play a role in cell growth, differentiation, and overall regulation of cell function.

Hern and Hubbell showed that the incorporation of the adhesion peptide RGD into a non-adhesive hydrogel proved to be useful for tissue resurfacing. There are numerous other adhesion peptides for targeting particularly desirable cell types and to modulate biological responses. Urokinase has been widely used for the clinical treatment of thrombogenetic disease and hemorrhoidal disease.

Artificial organ materials, on which urokinase was immobilized for its fibrinolytic activity, have been developed for blood-compatible materials. Immobilized urokinase by encapsulation in poly(2-hydroxyethyl methacrylate) and Introduced urokinase on the surface of the polytetrafluoroethylene using plasma modification technique by covalent bond. Another example of immobilized urokinase application was reported by Kato and coworkers, who had used urokinase immobilized in a Teflon catheter for treatment of thrombosis. Most of the studies found in the literature, regarding the surface functionalization of biomaterials with biological molecules, include the incorporation of adhesion and differentiation factors. The same approach may be used to incorporate specific enzymes able to regulate a number of cell functions. For instance, it is known that mitogen activated protein (MAP) kinase, upon activation by dual phosphorylation at threonine and tyrosine residues, is able to activate downstream targets that have been implicated in controlling gene expression, cell differentiation, and proliferation. This enzyme may be immobilized on the surface of biomaterials to control cell response, but other possibilities using different enzymes remain unexplored.

Enzyme Delivery

Although purified enzymes are now available for some enzyme deficiency diseases there are many problems in delivering the enzyme to the required site under such conditions that it will remain stable and active for a reasonable time. Normally, quite large amounts of enzyme are necessary with high level of purity and in a non-immunogenic form. In addition, many enzymes when administered are inactivated or degraded fairly rapidly. The delivery of therapeutic molecules requires, therefore, efficient strategies to have a precise control on their release profile according to specific locations.

It might be possible to control the release of such molecules by creating delivery systems sensitive to changes in pH, temperature, or salt concentration or to the feedback provided by cells. The concept of enzyme-activated drugs in therapy is scientifically, as well as clinically, attractive, as it allows the chemist and enzymologist full intellectual rein in designing interlinked systems. As therapeutic drugs, enzymes possess several attributes such as high specificity towards substrate, high solubility for preparing liquid formulations, and optimum activity under physiological conditions. The administration of enzymes, in cases of enzyme deficiency and inborn errors of metabolism and in the treatment of certain types of cancer, appears to offer a successful form of therapy.

Cancer therapy based on the delivery of enzymes to tumor sites has advanced in several directions since antibody-directed enzyme/prodrug therapy was first described. Nanospheres, nanocapsules, liposomes, micelles, and other nanoparticulates are frequently referred to as carriers for delivery of therapeutic and diagnostics agents. Asparagine is an essential amino acid for certain types of leukemias that lack asparagines synthetase activity. The activity of *L* - asparaginase is to degrade asparagine into aspartate and ammonia. Therefore, asparaginase has been of interest to biochemists and clinicians as a possible cancer therapeutic agent. Some success has been achieved in administering asparaginase in capsules made of nylon and polyurea to mice and rats.

Although asparaginase has been found to be effective in the treatment of some patients, it may have several serious side effects. Relatively high concentrations of the enzyme are needed for it to be clinically effective. These levels cause a wide range of toxic effects on several organs including the liver, pancreas, kidneys, and brain. The enzyme may also be recognized as foreign by the body and potentially severe immunogenic responses will be stimulated, resulting in hypersensitivity reactions.

To overcome these problems, immobilized enzyme derivatives have been prepared on various supports for extracorporeal treatment. Enzymes may be also used in cancer therapy as prodrug activators.

This therapy consists basically in using a drug that has been chemically modified so that it remains inactive until specifically activated by an enzyme at the target site. Another example of an anticancer enzyme is hyaluronidase. Hyaluronidase is a globular enzyme of endoglycosidase action, which can depolymerize hyaluronic acid in the organism, decreasing its viscosity and increasing tissue permeability.

Hyaluronidase has been utilized extensively as an adjunct in anticancer chemotherapy regimens, suggesting that hyaluronidase has intrinsic anticancer properties against tumor growth. *In vitro* studies in tissue culture with tumor spheroids and *in vivo* tests using animal models demonstrated the beneficial effect of hyaluronidase for the penetration of drugs into tumor tissue. Later, in a prospective clinical trial, hyaluronidase significantly improved the outcomes of patients with bladder carcinoma if the enzyme was administrated topically together with mitomycin.

Hyaluronidase is also used for local application (subcutaneous injections) during treatment of joint disease, in dermatology, and in ophthalmology. Hyaluronidase had been used in ophthalmology with the aim of formation of a thinner scar and to prevent necrosis after paravataes with zytostatics.

Trypsin has also been used to remove dead tissue from wounds, burns, and ulcers to speed the growth of new tissue and skin grafts, as well as to inhibit the growth of some contaminant organisms. The inappropriate activation of trypsinogen within the pancreas leads to development of pancreatitis. Once trypsin is activated, it is capable of activating many other digestive pro-enzymes.

These activated pancreatic enzymes further enhance the auto-digestion of the pancreas. Many materials, such as nylon, polysulfone, glycidyl methacrylate, chitosan, cellulose, and cellulose derivatives, have been used for trypsin immobilization. Enzyme therapy has also been tested for pancreatic insufficiency and cystic fibrosis. Pancreatic insufficiency can be alleviated by administrating orally enteric-coated microspheres containing lipase, amylase, and proteases. A special polymer coating protects the enzymes at low pH, such as in the stomach, and then releases them in the intestine at physiological pH. Lysozyme is a good example of an enzyme that catalyzes chemical reactions in the cell.

Lysozyme acts to kill bacteria by cleaving the covalent bond between the alternating polysaccharides that compose peptidoglycan in bacterial cell walls. The human salivary defence proteins and lysozyme are known to exert a wide antimicrobial activity against a number of bacterial, viral, and fungal pathogens *in vitro*. Therefore, these proteins, alone or in combinations, have been incorporated as preservatives in foods and pharmaceuticals as well as in oral health care products to restore saliva's own antimicrobial capacity in patients with dry mouth. These antimicrobials used in oral health care products, such as dentifrices, mouth rinses, moisturizing gels, and chewing gums, have been purified from bovine colostrum.

Other studies had been reported with lysozyme bound to chitosan, silica gel by means of physical adsorption, crosslinking to a polystyrene divinylbenzene matrix by the formation of ionic bindings, and by covalent attachment to non-porous glass beads. Harada and Kataoka described lysozyme immobilized into poly(ethylene glycol)–poly(aspartic acid) micelle. Lysozyme was selected as a model protein to incorporate into the micelle because it has a high isoelectric point (pI = 11), is positively charged over a wide range of pH, and has practical usage in drug delivery application as a lytic enzyme. Chen and Chen prepared immobilized lysozyme by carbodiimide method to form amide bonds with an enteric coating polymer (hydroxypropyl methyl-cellulose acetate succinate [AS-L]) as the carrier, which shows reversibly soluble-insoluble characteristics with pH changes.

The glucose-6-phosphate dehydrogenase enzyme catalyzes the oxidation of glucose-6-phosphate to 6-phosphogluconate while concomitantly reducing the oxidized form of nicotinamide adenine dinucleotide phosphate (NADP+) to nicotinamide adenine dinucleotide phosphate (NADPH). NADPH, a required cofactor in many biosynthetic reactions, maintains glutathione in its reduced form. Reduced glutathione acts as a scavenger for dangerous oxidative metabolites in the cell. With the help of the enzyme glutathione peroxidase, reduced glutathione also converts harmful hydrogen peroxide to water. Red blood cells rely heavily on glucose-6-phosphate dehydrogenase activity because it is the only source of NADPH that protects the cells against oxidative stresses. People deficient in glucose-6-phosphate dehydrogenase are not prescribed, therefore, with oxidative drugs because their red blood cells undergo rapid hemolysis under this stress.

In Greece, glucose-6-phosphate dehydrogenase deficiency is the main cause of severe neo-natal jaundice. The deficiency of this enzyme affects all races; the highest prevalence is among persons of African, Asian, or Mediterranean descent. Study of immobilized glucose-6-phosphate dehydrogenase has been reported. Kotorman *et al.* immobilized glucose-6-phosphate dehydrogenase from yeast on polyacrylamide beads possessing carboxylic functional groups activated by a watersoluble carbodiimide.

They verified highest operational stability of immobilized glucose-6-phosphate dehydrogenase. In relation to the use of immobilized enzymes with therapeutic purposes, catalase is one of the most interesting because it is employed to accelerate healing as well as to correct hereditary deficiencies and, in combination with hydrogen peroxide, as an antiseptic against anaerobes. The catalase enzyme has the ability to decompose hydrogen peroxide into oxygen and water, playing a central role in controlling the hydrogen peroxide concentration in human cells. More than 98 per cent of blood catalase is localized in erythtocytes.

These cells, with their high catalase level, provide a general protection against the toxic concentration of this small hydrogen peroxide molecule. The deficiency of catalase could cause acatalasemia. Several methods have been developed for the immobilization of catalase. Immobilization is often accompanied by changes in the enzymatic activity, optimum pH, affinity to the substrate, and stability. The extent of these changes depends on the enzyme, carrier support, and the immobilization conditions. The shift in the optimum pH, from acidic or alkaline to neutral pHs, may be useful for biomedical applications since it will allow the use of some enzymes (more active at low or high pHs) under more physiological conditions. The approach developed by Sakiyama-Elbert *et al.*, consisting in a cell-triggered growth factor delivery system, may also be used for the release of other important therapeutic molecules.

METAL ION CATALYSIS

Many enzymes require metal ions for maximal activity. If the enzyme binds the metal very tightly or requires the metal ion to maintain its stable, native state, it is referred to as a metalloenzyme. Enzymes that bind metal ions more weakly, perhaps only during the catalytic cycle, are referred to as metal activated. One role for metals in metal-activated enzymes and metalloenzymes is to act as electrophilic catalysts, stabilizing the increased electron density or negative charge that can develop during reactions. Among the enzymes that function in this manner is liver alcohol dehydrogenase. Another potential function of metal ions is to provide a powerful nucleophile at neutral pH. Coordination to a metal ion can increase the acidity of a nucleophile with an ionizable proton:

$$M^{2+} + NucH \rightleftharpoons M^{2+}(NucH) \rightleftharpoons M^{2+}(Nuc^{-}) + H^{+}$$

A metal such as Cu2+ or Zn2+ can also stabilize the TS. The metal must be able to be bound to the charged intermediate and hence the TS. The tetrahedral oxyanion intermediate of the reaction of an electrophilic carbonyl C can interact with a metal if there is an O on an adjacent atom which can help coordinate the metal ion.

This charge stabilization of the developing negative in the TS and the full negative in the intermediate is often called electrostatic catalysis. This method is likely to be found in many enzymes since nearly 1/3 of all enzymes require metal ions. A classic example of an enzyme using metal ion catalysis is carboxypeptidase A.

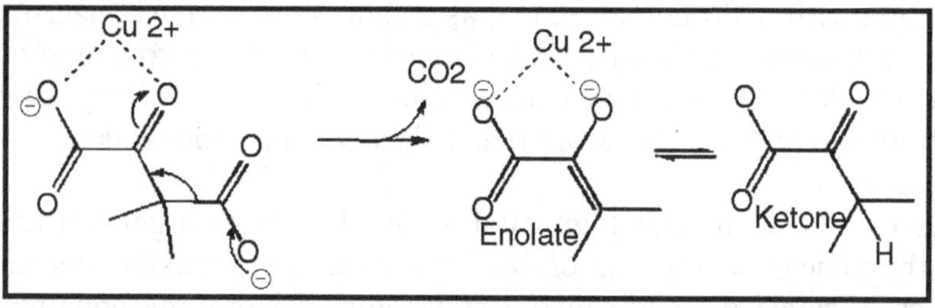

Fig. Metal Ion Or Electrostatic Catalysis - Stabilization Of Ts Charge

Metals can also act in a different way. They may coordinate a water and by further polarizing the H-O bond increase the acidity of the bound water. For instance, the water molecule in the pentammineaquacobalt(III) ion has a pKa of 6.6, compared to pure water, with a pKa of 15.7.

(To calculate the latter, write the equilibrium expression for water: H2O + H2O <==> H3O+ + OH-. Then write the Ka expression, as for a generic acid, which is [H3O+][OH-]/[H2O] = 10-14/55.5. The pKa is 15.7). The complexed hydroxide is a better nucleophile than bulk water. An example of an enzyme whose bound metal increases the nucleophilicity of water is carbonic anhydrase.

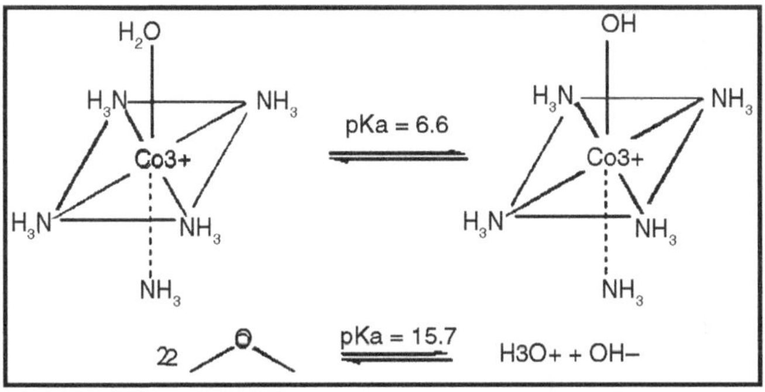

Fig. Metal Ion Catalysis - Decreasing Pka Of Water

COVALENT OR NUCLEOPHILIC CATALYSIS

One way to change the activation energy of the reaction is to change the reaction mechanism in ways which introduces new steps with lower activation energy. A typical way is to add a nucleophilic catalyst which forms a covalent intermediate with the reactant. The original nucleophile can then interact with the intermediate in a nucleophilic substitution reaction. If the nucleophilic catalyst is a better nucleophile than the original nucleophile (usually water) then the reaction is catalyzed. The nucleophilic catalyst and the original nucleophile usually interact with a carbonyl C in a substitution reaction, initially forming the tetrahedral oxyanion intermediate.

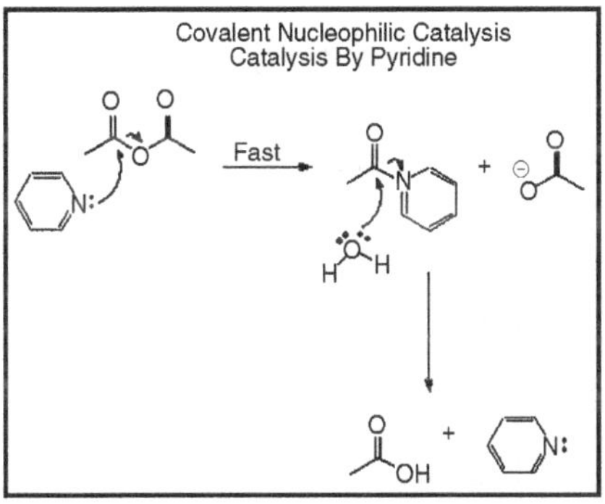

Fig. Nucleophilic Covalent Catalysis By Pyridine

If an amine is used as the nucleophilic catalyst, then the initial addition product (a carbinolamine) can become dehydrated, since the free pair of electrons on the N are more likely to be shared with the carbon to form a double bond than electrons from the original carbonyl O, which is more electronegative than the N). An imine or Schiff Base forms, with a pKa of about 7. This is easily protonated to form a positively charged N at the former carbonyl O centre. This serves as an excellent electron sink for decarboxylation reactions of beta-keto acids and illustrates an important point. Electrons in chemical reactions can be viewed as flowing from a source (such as a carboxyl group) to a sink.

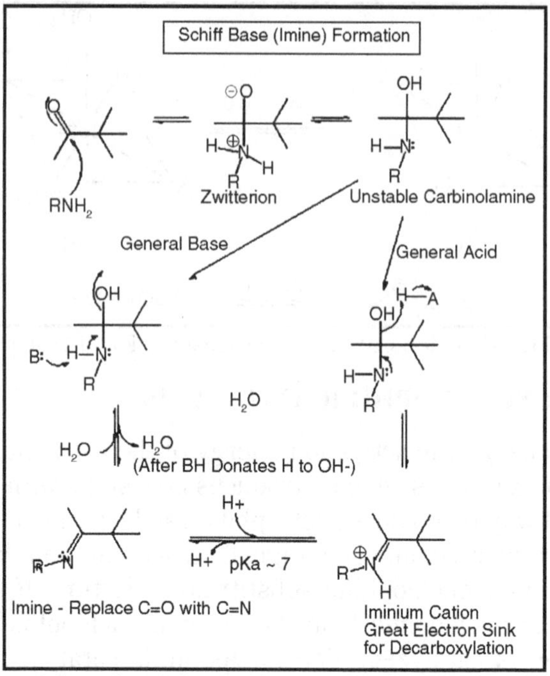

Fig. Mechanism Of Nucleophilic Catalysis by Amines - Schiff Base Formation.

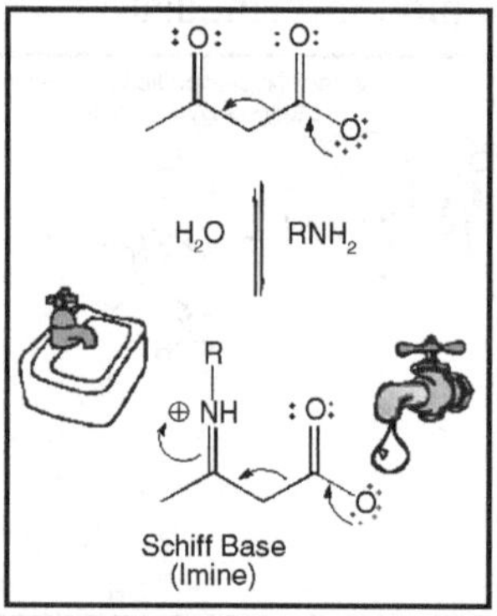

Fig. Electron Flow: Source To Sink

An intriguing question arises: how much of the structure of a large protein is really needed for catalysis? Much work has been directed to the development of small molecule catalysis mimetics of large protein enzymes. Just how small can you go in reducing the size of a protein and still get catalysis. One important feature of enzyme catalysis is that they catalyze reactions in which only one enantiomer is produced. That is, the synthesis is assymertric. This is typically a consequence of the asymmetric enzyme (itself chiral) binding only one enantiomer as a reactant and/or the imposition of steric restrictions on the possible reactions of the bound substrate. Recently, it has been show that L-Pro alone can act as such an assymetric catalyst in an aldol condensation reaction.

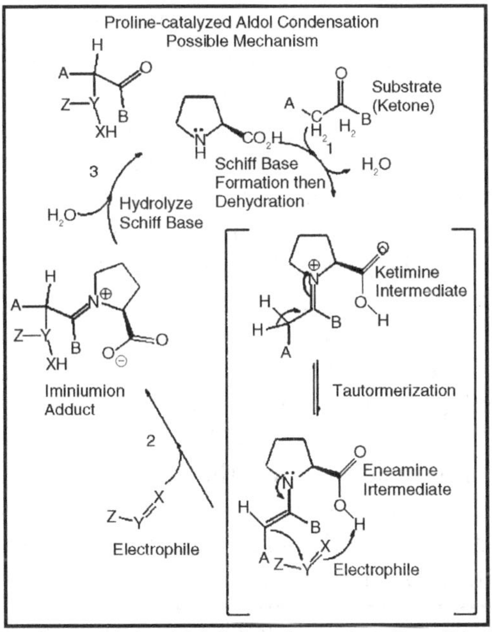

Fig.L-Pro Catalysis Of An Aldol Condensation: Possible Mechanism

INTRAMOLECULAR CATALYSIS

Consider the hydrolysis of phenylacetate. This reaction, a nucleophilic subsitution reaction, could be catalyzed by the addition to solution of the general base acetate, as described above. Since this reaction would double with the doubling of the solution acetate, the reaction is bimolecular (first order in reactant and catalyst). Now consider the same reaction only when the the general base part of the catalyst, the carboxyl group, is part of the reactant phenylacetate.

Such a case occurs in the acetylated form of salicylic acid - *i.e.* aspirin. When the carboxy group is ortho compared to the acetylated phenolic OH, it is in perfect position to accept a proton from water, decreasing the charge development on the O in the transition state. The general base does not have to diffuse to the appropriate site when it is intramolecular with respect to the carbonyl C of the ester link.

The rate of this intramolecular base catalysis is about 100 fold greater than of an intermolecular base catalyst like acetate. It is as if the effective concentration of the intramolecular carboxyl base catalyst is much higher duc to its proximity to the reaction site. Another type of reactions involving a carboxyl group (in addition to simple proton transfer) is when the negatively charged carboxyl O acts as a nucleophile and attacks an electrophilic carbonyl carbon. When the carbonyl is part of an ester, the carboxyl group engages in a nucleophilic substitution reaction, expelling the alcohol part of the ester as a leaving group.

The remaining examples below consider the nucleophilic (carboxyl) substitution on phenylesters, with phenolate as the leaving group. The reactions in effect transfer an acyl group to the carboxyl group to create an anhydride. First consider acyl transfer with aspirin derivatives. Aspirin, as you know, contains a carboxyl group ortho to an ester substitutent. Hence the carboxyl group can act as a nucleophile and attack the carbonyl carbon of the ester in a nucleophilic substitution reaction. The net effect is to transfer the acetyl group from the phenolic OH to the carboxyl group converting it to an anhydride. This is an intramolecular reaction. Compare this reaction to a a comparable bimolecular reaction shown below.

The first order rate constant of the intramolecular transfer of the acetyl group to the carboxyl group, $k_1 = 0.02\ s^{-1}$. The analogous bimolecular reaction rate constant $k_2 \sim 10^{-10}\ M^{-1}s^{-1}$. Dividing k_1/k_2 gives the relative rate enhancement of the intramolecular over the intermolecular reaction. With units of molarity, this ratio can be interpreted as the relative effective concentration of the intramolecular nucleophile. This makes the effective concentration of the carboxylate in the aspirin derivative 2×10^7 M.

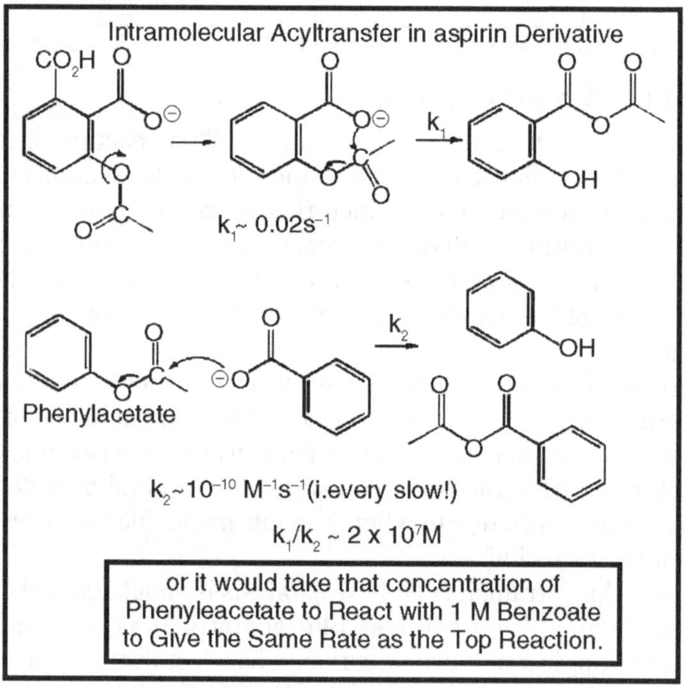

Fig. Acyl Transfer In Aspirin Derivatives.

Now consider the cleavage of phenylacetate using acetate as the nucleophile. The products are acetic anhydride and phenolate. This is a bimolecular reaction (a slow one at that), with a bimolecular rate constant, k_2 which I will arbitrarily set to 1 for comparison to some similar reactions.

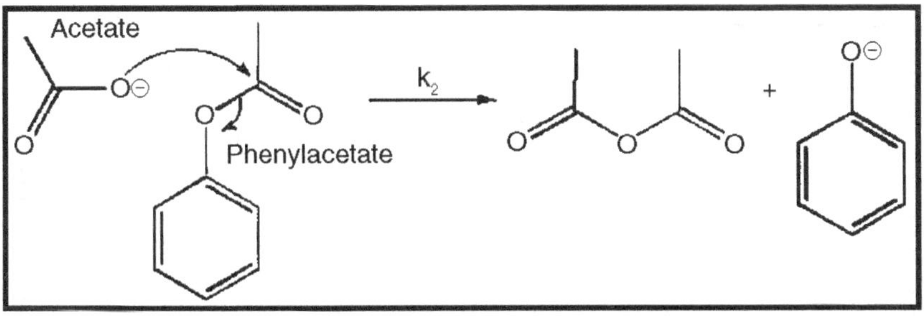

Fig. Reaction of Acetate With Phenylacetate

Now consider a monoester derivatives of succinic acid - phenyl succinate - in which the free carboxyl group of the ester attacks the carbonyl carbon of the ester derivative.

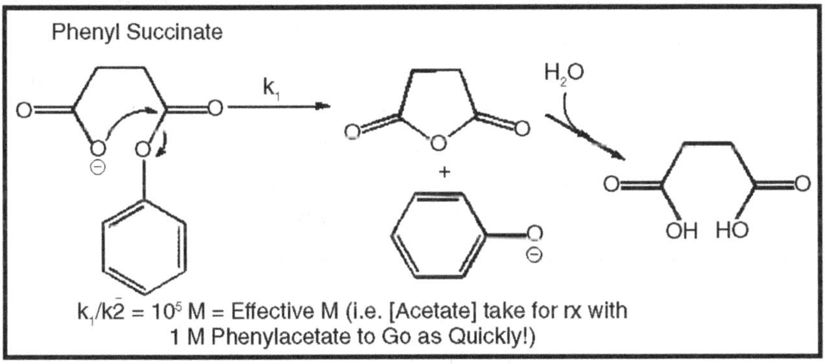

Fig. Intramolecular Reaction of Phenylsucciate.

If you assign a second order rate constant $k_2 = 1\ M^{-1}s^{-1}$ to the analogous intermolecular reaction of acetate with phenylacetate (as described above), the first order rate constant for the intramolecular reaction of phenylsuccinate is $10^5\ s^{-1}$.

The ratio of rate constants, $k_1/k_2 = 10^5$ M. That is it would take 10^5 M concentration of acetate reacting with 1 M phenylacetate in the first bimolecular reaction to get a reaction as fast as the intramolecular reaction of phenylsuccinate. An even more sterically restricated bicyclic phenylcarboxylate shows a $k_1/k_2 = 10^8$ M.

Another example is anhydride formation between two carboxyl groups. The DG° for such a reaction is positive, suggesting an unfavourable reaction. Consider two acetic acid molecules condensing to form acetic anhydride.

For this intermolecular reaction, $K_{eq} = 3x10^{-12}\ M^{-1}$. Now consider the analogous intramolecular reaction of the dicarboxylic acid succinic acid. It condenses in an intramolecular reaction to form succinic anhydride with a $K_{eq} = 8x10^{-7}$ (no units). The ratio $K_{eq\text{-}intra}/K_{eq\ inter} = 3 \times 10^5$ M. It is as if the effective concentration of the reacting groups. because they do not have to diffuse together to react, is 3 x 10^5 M.

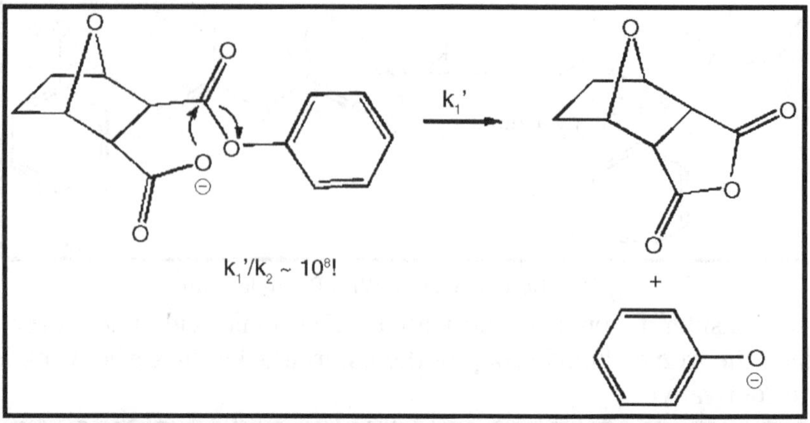

Fig. Intramolecular Reaction of Bicyclic Phenylcarboxylate

How does this apply to enzyme catalyzed reaction? Enzymes bind substrates in physical steps which are typically fast. The slow step is chemical conversion of the bound substrate, which is effectively intramolecular. These three kinds of reactions, intermolecular, intramolecular, and enzyme-catalysed can be broken down into two hypothetical steps, a binding followed by catalysis.

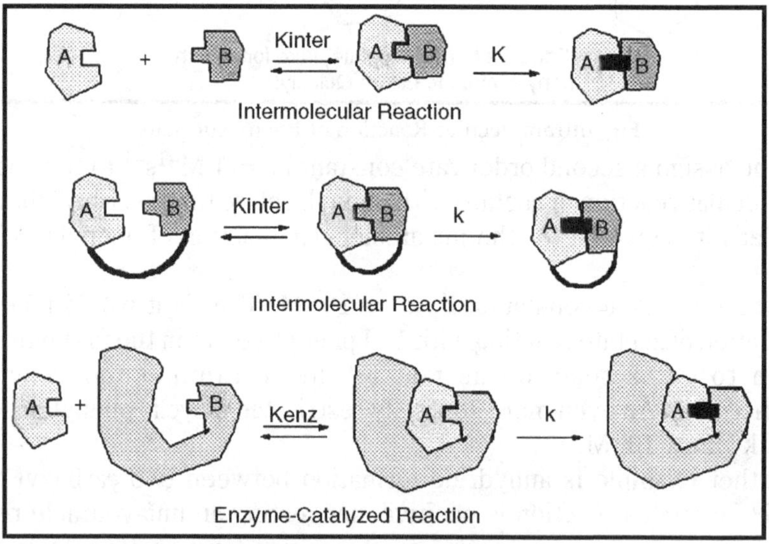

Fig. Three kinds of reactions, Intermolecular, Intramolecular, and Enzyme-Catalysed

If the rate constants for the chemical steps are all identical, the advantage of the intramolecular and enzyme-catalyzed reaction over the intermolecular reaction is K_{INTRA}/K_{INTER} and K_{ENZ}/K_{INTER}, respectively.

Calcium in solution exists as a octahedrally coordinated complex with water occupying all the coordination sites. EDTA, a multidentate ligand, first interacts through one of its potential six electron donors to Ca in a reaction which is entropically disfavoured from the the Ca-EDTA perspective, although one water is released. Once this first intramolecular complex is formed, the rest of the ligands on the EDTA rapidly coordinate with the Ca and release bound water. The former is no longer entropically disfavoured since it is now an intramolecular process while the later is favoured through the release of the remaining five water molecules.

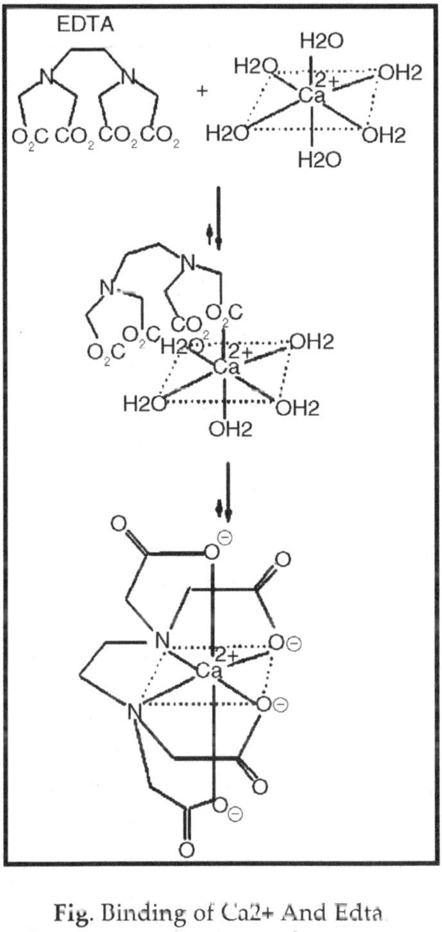

Fig. Binding of Ca2+ And Edta

We modeled the catalytic advantage offered by intramolecular reaction in terms of a dramatic increase in the effective concentration of reactants, which sometimes reached levels of 10^8 M Another way is to look at entropy changes associated with dimer formation.

TRANSITION STATE STABILIZATION

Linus Pauling postulated long ago that the only thing that a catalyst must do is bind the transition state more tightly than the substrate. That this must be the case can be seen from the diagram below, which shows how S and S* (the transition state) can react with E to form a complex which then proceeds to product, or can go to product in the absence of E.

From this diagram, it should be evident that c – a = d –b, where a is the DGº for the binding of S to E, and b is the DGº for the binding of S* to E. For an enzyme to be a catalyst the activation energy for the reaction in the presence of E, d, must be less than in the absence of enzyme, c. Therefore c-d = a-b > 0. Since DGº = -RTln K_{eq}, K_{eq} for binding of S* to E is greater than for S binding to E.

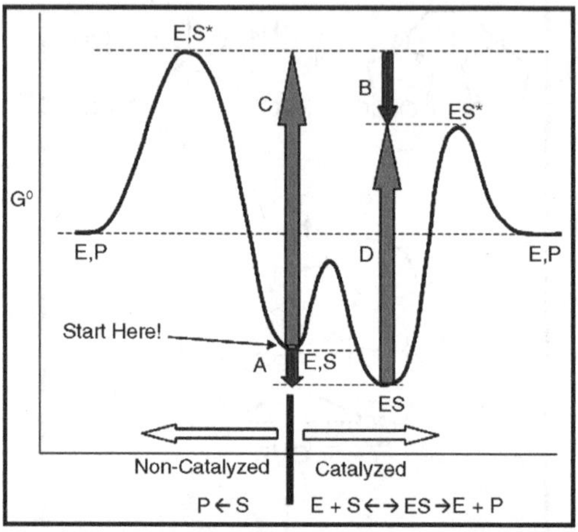

Fig. Enzymes Bind The Transition State More Tightly than The Substrate

The stability of the transition state also affects the reaction kinetics (which makes sense given that the activation energy clearly affects the speed of a reaction). As you probably remember from organic chemistry, S_N2 reactions are slow when the central atom where the substitution will occur is surrounded by bulky substitutents.

(Sterics once again.) We discussed this in context to nucleophiliic substitution on a sp^2 hybridized carbonyl carbon in carboxylic acid derivatives versus on a sp^3 hybridized phosphorous in phosphoesters and diesters. The explanation for this phenomena has usually been attributed to hindered access of the central atom caused by bulky substituents (intrinsic effects).

Is this true? Recent studies on S_N2 reactions of methylchl-oroacetonitrile and t-butylchloroacetonitrile (with the reagent labeled with ^{35}Cl) using $^{37}Cl^-$ as the incoming nucleophile in the gas phase shown that the more hindered t-butyl derivative's activation energy was only 1.6 kcal/mol higher than the methyl derivative, but in aqueous solution, the difference is much greater for comparable reactions. They attributed the differences to solvation effects of the transition state. The bulkier the substituents on the central atom, the more difficult it is to solvate the transition state since water can't reorient around it as well. In effect there is steric hindrance for both reactant and solvent.

Abyzmes - Antibody Catalysis

What does it take for a macromolecule (M) to be a catalyst - an enzyme. *It seems the minimum criterion are*:

- M binds a reactant
- M binds the transition state more tightly than the substrate

Anything above these is just "icing on the cake". If different functional group are present in the "active" site of the enzyme that would allow electrostatic, intramolecular, covalent, general acid and/or base catalysis, the better the catalyst.

Linus Pauling recognized the two key factors decades ago. He made the following hypothesis: Antibody molecules (immune system proteins that bind foreign molecules) that can be made to bind to transition state analogs of a substrate, should also presumably catalyze the conversion of substrate, through the transition state, to product.

About a decade ago, his prediction was verified. Lerner *et al.* made a transition state analog of an ester. When an ester is hydrolyzed, the sp^2 hybridized carbonyl carbon is converted to an sp^3 hybridized centre in the intermediate, with the carbonyl oxygen becoming an oxyanion.

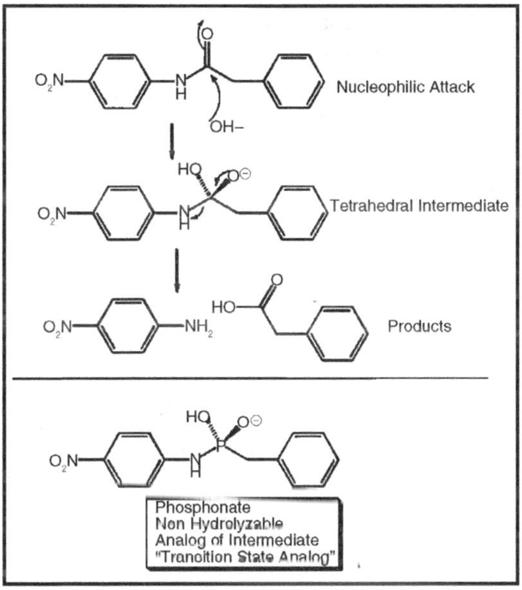

Fig. Phosphonates: Transition State Analogs

The transition state presumably looks more like this unstable intermediate (sp^3, oxyanion). Lerner synthesized a phosphonate, an ester mimic with a sp^3 hybridized phosphorous replacing the carbonyl C. It also has a negatively charged oxygen as does the intermediate for the ester.

This phosphonate ester is very resistant to hydrolysis. When injected into a mouse (after first being covalently attached to a carrier protein so the small molecule becomes "immunogencic"), the mouse makes a protein antibody which binds to the phosphonate.

When the corresponding carboxylic acid ester is added to the antibody, it is cleaved with nominal k_{cat} and Km values. Site specific mutagenesis can then be done to make it an even better catalyst! The antibody enzymes have been called abzymes.

PROXIMITY

Chemical reactions go faster when the reactants are in proximity, that is, near each other. In solution or in the gas phase, this means that increasing the concentrations of reacting molecules, which raises the number of collisions, causes higher rates of reaction.

Enzymes, which have specific binding sites for particular reacting molecules, essentially take the reactants out of dilute solution and hold them close to each other.

This proximity of reactants is said to raise the "effective" concentration over that of the substrates in solution, and leads to an increased reaction rate. In order to measure proximity effects in enzyme reactions, enzymologists have turned to model studies comparing intermolecular reaction rates with corresponding or similar intramolecular reaction rates.

A typical case is the imidazole-catalyzed hydrolysis of p-nitrophenylacetate. Under certain conditions the rate constant for this bimolecular reaction is 35 M-1 min-1. By comparison, the first-order rate constant for the analogous but intramolecular reaction is 839 min^{-1}.

The ratio of these two rate constants, has the units of concentration and can be thought of as an effective concentration of imidazole in the intramolecular reaction. Put another way, a concentration of imidazole of 23.9 M would be required in the intermolecular reaction to make it proceed as fast as the intramolecular reaction.

$$(839\ min^{-1})/(35\ M^{-1}\ min^{-1}) = 23.97\ M$$

There is more to this story, however. Enzymes not only bring substrates and catalytic groups close together, they orient them in a manner suitable for catalysis as well.

Comparison of the rates of reaction of the molecules makes it clear that the bulky methyl groups force an orientation on the alkyl carboxylate and the aromatic hydroxyl groups that makes them approximately 250 billion times more likely to react.

Enzymes function similarly by placing catalytically functional groups (from the protein side chains or from another substrate) in the proper position for reaction.

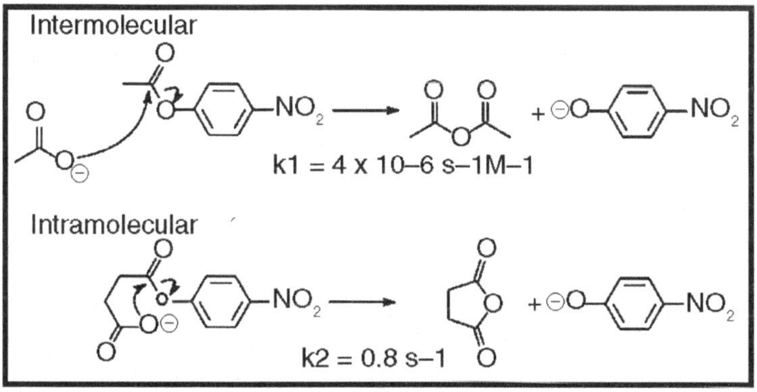

Fig. Proximity Effects in Catalysis

Clearly, proximity and orientation play a role in enzyme catalysis, but there is a problem with each of the above comparisons. In both cases, it is impossible to separate true proximity and orientation effects from the effects of entropy loss when molecules are brought togetherare much smaller than the values. Simple theories based on probability and nearest-neighbour models, for example, predict that proximity effects may actually provide rate increases of only 5- to 10-fold. For any real case of enzymatic catalysis, it is nonetheless important to remember that proximity and orientation effects are significant.

RNA AND ANTIBODY MOLECULES AS ENZYMES: RIBOZYMES AND ABZYMES

RNA AND DNA AS CATALYSTS

DNA and RNA are usually seen as carriers of genetic information, with the occasional function of RNA as a scaffold, as for example in the ribosome. This picture has changed with the discovery of self-splicing of certain RNAs, most notably that of the Tetrahymena group I intron by Tom Cech, and the active role of RNA in the RNase P in the process of maturation of tRNAs by Sid Altman. These observations laid the foundation for the concept of catalytic RNA for which the Nobel prize of 1989 was awarded to these two colleagues.

Natural catalytic RNAs

These two first examples of catalytic RNA have been joined by others such as the group II intron, the hepatitis delta RNA and the plant viroid RNAs of the hammerhead and hairpin type. Although some of these reactions are facilitated by proteins *in vivo*, their reactions can be simulated *in vitro* without the aid of proteins, supporting the view that indeed the RNA by itself can do the job. With the exception of RNase P RNA, the RNAs undergo an intramolecular, or in *cis*, transesterification in these transformations.

These reactions are sequence specific, and more importantly, occur with rates orders of magnitude faster than expected from the reactivity of RNA.

Thus there must be a catalytic element involved, and therefore these active RNAs are called ribozymes, in analogy to the classical protein enzymes. To study the mechanism of these ribozyme-catalysed reactions and their structural requirements, they were redesigned for intermolecular, or *in trans*, reactions. Thus substrate and ribozyme are localised on different RNA strands. This arrangement facilitates the determination of kinetic constants such as kcat and Km values.

An example for an *in trans* arrangement is shown for the hammerhead ribozyme. A comparison of the efficiency of ribozyme- with enzyme-catalysed reactions shows that even though the reaction rate kcat is often smaller for the ribozymes, they compensate for it by a more favourable Km so that in the end efficiency is comparable for both catalysts.

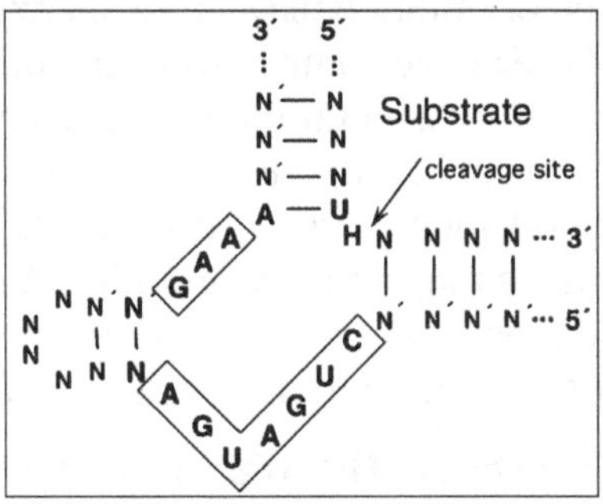

Fig. Generalised Two Dimensional Structure of the Hammerhead Ribozyme for Cleavage in Trans.

N, any nucleotide; N´, nucleotide complementary to N; H, any nucleotide except G; underlined, invariant core.

Mechanistic Aspects

Most of these RNA-catalysed reactions depend on the presence of a divalent metal ion such as Mg^{2+} but some, such as the hairpin and the hammerhead ribozyme, can also be coaxed into action with high concentrations of monovalent ions such as Na^{+}. Very interestingly, certain positively charged aminoglycosides can replace the metal ions for the hairpin ribozyme. Thus the question as to the role of the metal ions arises. Formation of a kinetically competent three-dimensional structure is quite certainly one of them. However, particularly in the presence of Mg^{2+} where most ribozymes work best, a more active role can be envisaged where it acts as a base and as a Lewis acid as originally proposed in the "two metal ion" mechanism for enzymes. Such mechanisms have been formulated particularly for the group I RNA and the hammerhead ribozymes.

In these cases, one metal ion is thought to abstract a proton from the nucleophilic hydroxy group to attack the phosphate group, and the other coordinates to the departing hydroxy group. Although there is experimental support for such a metal ion hydrate role, no metal ions which could take on this function have been detected in x-ray structures of these ribozymes. Thus, the final proof for the involvement of the metal ions as active participants in the chemical step of the reactions is still lacking.

This of course becomes a concern in light of some of the ribozymes not requiring metal ions. Looking for alternative functional groups one has to consider whether the nucleobases could not play such a role. The argument against this idea in the past has been that the pKa values of these groups are outside the pH optimum of the reactions. NMR studies have however shown that the pKa of adenosine, for example, can be shifted upwards by 2.5 units to approximately 6.5 as a consequence of a change in the micro-environment in the three-dimensional structure of the ribozymes. Thus it might well be that such groups can function as acid-base catalysts at neutral pH and that we have to reconsider our prejudice against nucleotides as active participants in such catalytic reactions.

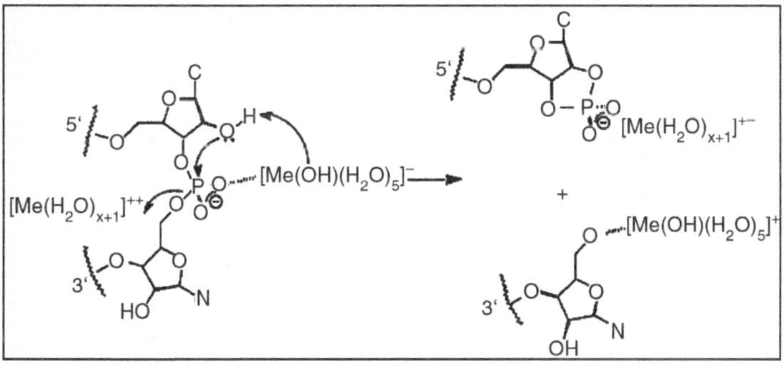

Fig. Suggested Hammerhead Ribozyme Cleavage Mechanism with Participation of Two Metal Ions.

New Ribozymes

The catalytic power of the naturally occuring ribozymes which all catalyse phosphoryl tranfers within RNA molecules, has stimulated the search for ribozymes with other activities. This has been made possible by the technique of *in vitro* selection. It involves the construction of double-stranded DNA containing a T7 promoter followed by a stretch of randomised nucleotides. Sequences necessary for binding of oligonucleotides as polymerase primers and for cloning are at both ends of such constructs. After transcription the RNA is incubated with the substrate and active RNA is separated from inactive RNA sequences. The success of the procedure depends crucially on the design of this selection step.

Often substrates with a biotin moiety are used which is transferred to the active RNA by the catalysed reaction. This facilitates separation of the active from the inactive RNA on a Streptavidin column. The active species are amplified by PCR, transcribed and incubated with the substrate again. This cycle is repeated between 6 and a dozen times until enough active RNA has accumulated for characterisation.. The method is extremely powerful, enabling the selection of active molecules from a large pool of sequences. Thus, for example, a DNA with 22 randomised positions has 4^{22} sequences which is equivalent to 1.7×10^{13} sequences. This represents a rather small degree of randomisation in the generally performed selection procedures. However, there are certain constraints created by Avogadro´s number, i. e. the number of molecules in one mol which is 6×10^{23}.

It so happens that 39 randomisations is equivalent to 3.7×10^{23} sequences. Thus to cover the whole sequence space one would have to work with a 1 M DNA solution in one litre for the first selection. This becomes forbiddingly expensive in terms of enzyme required. Of course, larger degrees of randomisations can be used with the penalty that not the entire sequence space will be searched. Such selections have yielded a most impressive number of ribozymes, catalysing the most diverse reactions. Examples are phosphoryl transfer like a kinase reaction, nucleoside triphosphate polymerisation, acyl transfer, N-glycosidic bond formation to obtain a nucleotide, and formation of a carbon-carbon bond in a Diels-Alder reaction.

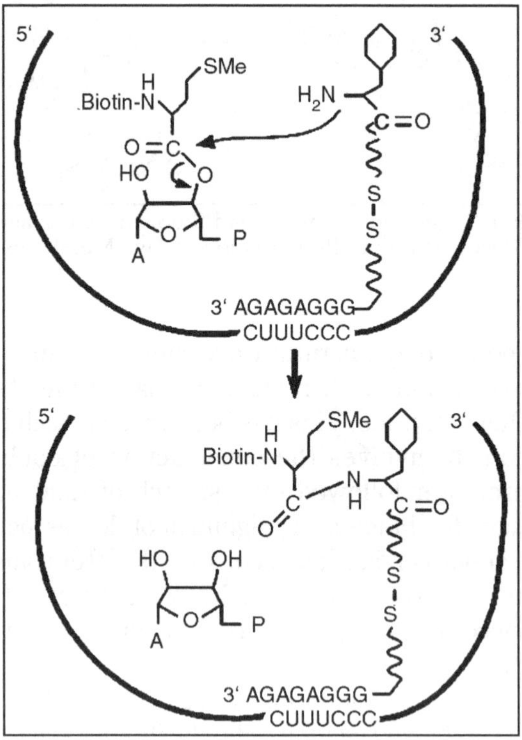

Fig. Example of an *in Vitro* Selected Ribozyme for Peptide bond Formation. Bold line, Ribozyme.

Surprisingly, the activity of ribozymes can be allosterically controlled. An example is the flavin-dependent activity of a hammerhead ribozyme to which a flavin-binding aptamer has been fused. In contrast to this positive effect, an ATP aptamer-ribozyme construct is inhibited in the presence of ATP. These examples obviously make ribozymes more akin to proteozymes. Besides the interest in mechanistic questions, such ribozymes might be very useful as therapeutic agents, as discussed below, as their activity can be modulated by small molecules.

DNAzymes

It was generally assumed that RNA was better suited as a catalyst than DNA because of its flexibility and the presence of the 2´-hydroxy group. It thus came as a big surprise when Joyce selected a DNAzyme capable of cleaving RNA, and this with somewhat better efficiency than a ribozyme. Thus our preconceived ideas have been proven wrong. DNA as a single-stranded molecule can apparently fold into three dimensional structures similarly to RNA, and the 2´-hydroxy group is dispensable for catalytic action.

As ribozymes, DNAzymes can also be made, by selection, to depend on a cofactor. This has been demonstrated for a histidine-dependent DNAzyme for RNA hydrolysis. This narrows the gap to proteozymes and offers the possibility of extending the range of catalysed reactions. So far, selections have been performed with the aim in mind of mimicking reactions which are also catalysed by proteozymes. Besides the interesting mechanistic aspects of the RNA- or DNA-catalysed reactions, the ribozyme-catalysed reactions particularly are often considered as support for the prebiotic RNA world where RNA might have played the part of our present day enzymes. As the number of various types of reactions catalysed by ribozymes increases, this hypothesis gains in strength. Another aspect for the future is the development, by *in vitro* selection, of nucleic acid enzymes to catalyse reactions for which no natural enzymes exist. This could be of industrial importance.

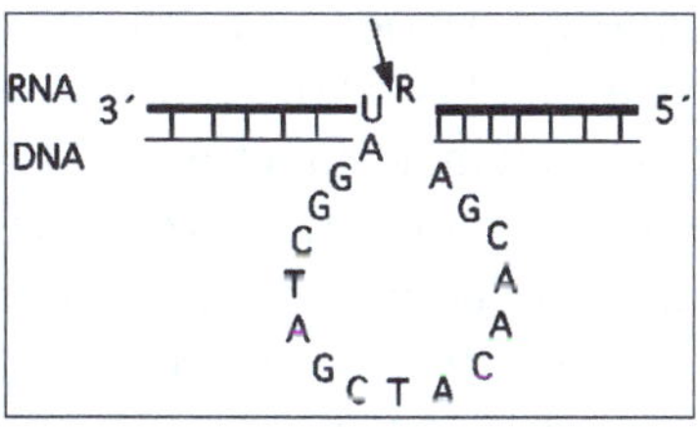

Fig. The first RNA-cleaving DNAzyme. R, Either Adenosine or Guanosine.

Catalytic Antibodies: Abzymes

Antibodies are *immunoglobulins*, which, of course, are proteins. Like other antibodies, catalytic antibodies, so-called abzymes, are elicited in an organism in response to immunological challenge by a foreign molecule called an antigen for discussions on the molecular basis of immunology). In this case, however, the antigen is purposefully engineered to be *an analog of the transition-state intermediate in a reaction.*

The rationale is that a protein specific for binding the transition-state intermediate of a reaction will promote entry of the normal reactant into the reactive, transition-state conformation. Thus, a catalytic antibody facilitates, or catalyzes, a reaction by forcing the conformation of its substrate in the direction of its transition state. (A prominent explanation for the remarkable catalytic power of conventional enzymes is their great affinity for the transition-state intermediates in the reactions they catalyze.

One strategy has been to prepare ester analogs by substituting a phosphorus atom for the carbon in the ester group. The phospho-compound mimics the natural transition state of ester hydrolysis, and antibodies elicited against these analogs act like enzymes in accelerating the rate of ester hydrolysis as much as 1000-fold.

Abzymes have been developed for a number of other classes of reactions, including COC bond formation via aldol condensation (the reverse of the aldolase reaction and the pyridoxal 5'-P-dependent aminotransferase reaction. In this latter instance, N_a-(5'-phosphopyridoxyl)-lysine coupled to a carrier protein served as the antigen.

An antibody raised against this antigen catalyzed the conversion of d-alanine and pyridoxal 5'-P to pyruvate and pyridoxamine 5'-P. This biotechnology offers the real possibility of creating "designer enzymes," specially tailored enzymes designed to carry out specific catalytic processes.

THE MECHANISM OF ENZYME CATALYSIS

In order for a reaction to occur, reactant molecules must contain sufficient energy to cross a potential energy barrier, the activation energy. All molecules possess varying amounts of energy depending, for example, on their recent collision history but, generally, only a few have sufficient energy for reaction. The lower the potential energy barrier to reaction, the more reactants have sufficient energy and, hence, the faster the reaction will occur.

All catalysts, including enzymes, function by forming a transition state, with the reactants, of lower free energy than would be found in the uncatalysed reaction. Even quite modest reductions in this potential energy barrier may produce large increases in the rate of reaction (*e.g.* the activation energy for the uncatalysed breakdown of hydrogen peroxide to oxygen and water is 76 kJ M^{-1} whereas, in the presence of the enzyme catalase, this is reduced to 30 kJ M^{-1} and the rate of reaction is increased by a factor of 10^8, sufficient to convert a reaction time measured in years into one measured in seconds).

The catalysed reaction pathway goes through the transition states TS_{c1}, TS_{c2} and TS_{c3}, with standard free energy of activation G_c^*, whereas the uncatalysed reaction goes through the transition state TS_u with standard free energy of activation G_u^*. In this example the rate limiting step would be the conversion of ES into EP. Reactions involving several substrates and products, or more intermediates, are even more complicated.

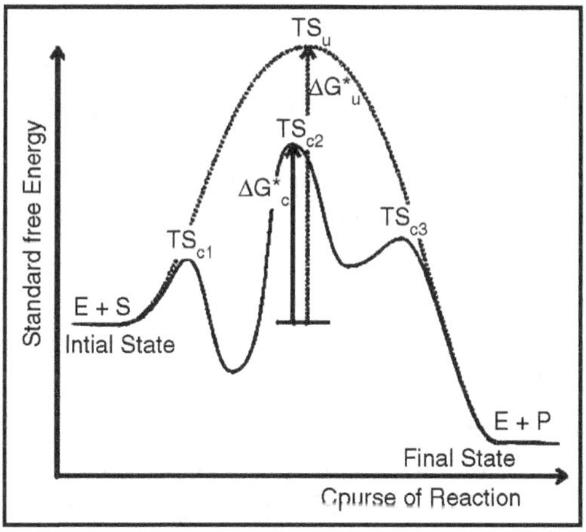

Fig. A Schematic Diagram Showing the free Energy Profile of the Course of an Enzyme Catalysed Reaction Involving the Formation of Enzyme-substrate (ES) and Enzyme-product (EP) Complexes, i.e.

$$E + S \rightleftharpoons ES \rightleftharpoons EP \rightleftharpoons E + P$$

The Michaelis-Menten reaction scheme would give a similar profile but without the EP-complex free energy trough. The schematic profile for the uncatalysed reaction is shown as the dashed line. It should be noted that the catalytic effect only concerns the lowering of the standard free energy of activation from G_u^* to G_c^* and has no effect on the overall free energy change (i.e.. the difference between the initial and final states) or the related equilibrium constant.

There are a number of mechanisms by which this activation energy decrease may be achieved. The most important of these involves the enzyme initially binding the substrate(s), in the correct orientation to react, close to the catalytic groups on the active enzyme complex and any other substrates. In this way the binding energy is used partially in order to reduce the contribution of the considerable activation entropy, due to the loss of the reactants' (and catalytic groups') translational and rotational entropy, towards the total activation energy. Other contributing factors are the introduction of strain into the reactants (allowing more binding energy to be available for the transition state), provision of an alternative reactive pathway and the desolvation of reacting and catalysing ionic groups.

The energies available to enzymes for binding their substrates are determined primarily by the complementarity of structures (i.e. a good 3-dimensional fit plus optimal non-covalent ionic and/or hydrogen bonding forces).

The specificity depends upon minimal steric repulsion, the absence of unsolvated or unpaired charges, and the presence of sufficient hydrogen bonds. These binding energies are capable of being quite large. As examples, antibody-antigen dissociation constants are characteristically near 10^{-8} M (free energy of binding is 46 kJ M^{-1}), ATP binds to myosin with a dissociation constant of 10^{-13} M (free energy of binding is 75 kJ M^{-1}) and biotin binds to avidin, a protein found in egg white, with a dissociation constant of 10^{-15} M (free energy of binding is 86 kJ M^{-1}).

However, enzymes do not use this potential binding energy simply in order to bind the substrate(s) and form stable long-lasting complexes. If this were to be the case, the formation of the transition state between ES and EP would involve an extremely large free energy change due to the breaking of these strong binding forces, and the rate of formation of products would be very slow.

They must use this binding energy for reducing the free energy of the transition state. This is generally achieved by increasing the binding to the transition state rather than the reactants and, in the process, introducing an energetic strain into the system and allowing more favourable interactions between the enzyme's catalytic groups and the reactants.

6

Chemical Changes and Microbes

At about this time Liebig the German chemist, had claimed that fermentation and putrefaction were processes caused by the chemical instability of organic molecules. On the other hand, Pasteur had shown that living yeast cells were required for the conversion of sugar into alcohol and had suggested that these fermentations and transformations are brought about by the agency of microorganisms.

At this time. the distilleries in France, where alcohol was produced from sugar beet had difficulties in their fermentations and requested Pasteur for help. Pasteur found that the trouble was due to the replacement of the alcoholic fermentation by a lactic fermentation. Instead of finding yeast in the fermenting brew, he found rod-shaped structures.

He concluded that these rod shaped organisms were responsible for the conversion of sugar into lactic acid. Subsequently, he also showed that different fermentations are brought about by specific microorganisms. For example, the alcoholic fermentation was the result of living yeast cells while the lactic acid fermentation was the result of the activity of lactic bacteria. Interestingly in 1897, after Pasteur, Buchner showed that intact yeast cells were not. required for the conversion of sugar into alcohol and carbon dioxide.

He prepared a cell free yeast juice by grinding yeast cells with sand and showed that this cell free juice had the ability to carry out the alcoholic fermentation. He concluded that the active agent in the yeast cells responsible for bringing about the fermentation was a protein and he named it as zymase. This finding gave support to Liebig's thesis that living cells were not required to bring about fermentations and marks the beginning of the study of enzyme chemistry.

ANCIENT MICROBIOLOGICAL HISTORY

Ancient man recognized many of the factors involved in disease. Early civilizations on crete, india, pakistan and scotland invented toilets and sewers; lavatories, dating around 2800 bc, have been found on the orkney islands and in homes in pakistan about the same time. One archaeologist has stated that "the high quality of the sanitary arrangements could well be envied in many parts of the world today". In rome, 315 ad, the public lavatories were places where people routinely socialized and conducted business. Ten to twenty people could be seated around a room, with their wastes being washed away by flowing water; it must have been difficult to "*stand on your dignity*" under such circumstances.

The chinese used toilet paper as early as ad 589. In europe moss, hay and straw were used for the same purpose. I can personally attest to the use as late as 1962 of "slick magazines" as toilet paper in certain european camp grounds. The first cities to use water pipes (of clay) were in the indus valley of pakistan around 2700 bc. Metal water pipes were used in egypt (2450 bc) and the palace of knossos on crete around 2000 bc had clay pipes.

Rome built elaborate aqueducts and public fountains throughout its empire to insure a clean supply of water for its citizens. Rome had a "water commissioner" who was responsible for seeing that the water supply was kept adequate and clean; the punishment for pollution of the water supply was death. Lead was commonly used for roman pipes and the subsequent fall of the roman empire has been related by some to the effects of lead on the roman brain.

Most ancient peoples recognized that some diseases were communicable and isolated individuals thought to carry "infections". An example of this is the universal shunning of lepers, which occurs even today. When the black death struck europe, entire villages were abandoned as people fled in an effort to escape the highly infectious plague.

Similarly, in the middle ages the rich of europe fled to their country homes when small pox struck in an effort to escape its terrible consequences. The fact that people who recovered from a particular disease were immune to that disease was probably recognized many different times in many places. Often these survivors were expected to nurse the ill. Greek and roman physicians routinely prescribed diet and exercise as a treatment for ills.

Sadly, we know that this knowledge did not help most of our ancestors and that the human life span was, until the last 200 years, more often than not cut short due to infectious disease. Even today approximately 15,000,000 children die per year, mainly from infectious diseases that are preventable with basic sanitation, immunization and simple medical treatments. One might honestly question just how far we have come in our treatment of disease. An excellent synopsis of the history of microbiology ancient people had certainly seen masses of microbes, such as mold and bacterial colonies, on spoiled food, but it is doubtful if anyone considered that they were viewing living organisms.

Small boys and maybe a few love-sick adults staring into a clear pond, must have seen tiny specks moving rapidly about and some may have considered them living creatures, but to express this to their friends would be equivalent to us telling our friends that we'd seen a flying saucer.

The first person to report seeing microbes under the microscope was an englishman, robert hooke. Working with a crude compound microscope he saw the cellular structure of plants around 1665. He also saw fungi which he drew. However, because his lens were of poor quality he was apparently unable to "see" bacteria.

Anton van leeuwenhoek was a man born before his time. Although not the first to discover the microscope or to use magnifying lens, he was the first to see and describe bacteria. We know that he was a "cloth merchant" living in delft holland. And that he used magnifying lens to view the quality of the weave of the merchandise he purchased. He traveled to england in 1668 to view english cloth and there he saw drawings of magnifications of cloth much greater than any of the current lens available in holland would do.

He returned to holland and took up lens grinding. Being meticulous, he developed his lens grinding to an art and in the process tested them by seeing how much detail he could observe with a given lens. One can guess that he chanced to look at a sample of pond water or other source rich in microbes and was amazed to see distinct, uniquely shaped organisms going, apparently purposefully, about their lives in a tiny microcosm. He made numerous microscopes from silver and gold and viewed everything he could including the scum on his teeth and his semen.

His best lens could magnify ~300-500 fold which allowed him to see microscopic algae and protozoa and larger bacteria. He clearly had excellent eyesight because he accurately drew pictures of microbes that were at the limit of the magnification of his lens. He used only single lens and not the compound lens of the true microscopes we employ today; which makes his observations all the more amazing. He wrote of his observations to the royal society of london in 1676 and included numerous drawings.

He astonished everyone by claiming that many of the tiny things he saw with his lens were alive because he saw them swimming purposefully about. This caused no end of shock and wonderment and numerous people hurried to delft to see if this dutchman was "in his cups" or if he was really onto something new and wonderful.

A few minutes with one of his numerous microscopes was all it took to convert his visitors to enthusiastic believers in the existence of these tiny beasties living all around them. His discovery was the equivalent of our finding life on mars today.

THE CLEAN NUT AND THE REVENGE OF THE BACTERIA

In the 1800s people began to use hospitals. Hospitals also became centers of physician training. In 1841 young doctor ignaz semmelweis was hired to run a maternity ward in a vienna hospital. There were two birthing wards in his preview, one run by midwives and the other by doctors. Semmelweiss noticed that the death rate among mothers in the doctor's ward ran as high as 18% from the blood infection known as child bed fever or puerperal sepsis, whereas in the midwife ward the death rate was much lower.

When he suggested that the doctors might contribute to this he was fired. Subsequently rehired, he saw a friend die from puerperal sepsis after cutting himself during an autopsy of a patient who had died of puerperal sepsis. He reasoned that there was an invisible agent that caused both deaths and that one could transfer it from the autopsy room to the birthing rooms and thus infect the mothers during birthing.

Acting on this assumption, semmelweis instituted sanitary measures which included having the doctors wash their hands in disinfectant and change from lab coats dripping with pus and blood from the autopsy room to clean lab coats before examining patients or assisting in a birth. The death rate of the mothers dropped by 2/3 in his ward. However, the other doctors objected so strongly to his rules that semmelweis was again fired and left vienna.

He took other hospital jobs where he instituted the same standards of cleanliness which resulted in the same decline in deaths and a revolt of his fellow physicians. Semmelweis ended up dying in an insane asylum from a blood infect that resembled puerperal sepsis.

SPONTANEOUS GENERATION

The mystery of life has puzzled and confounded humans since the first human began to contemplate his world. The religions of ancient societies were built around the seasons, the sun and the renewal of life as these were so clearly tied to survival; both of the human species through birth and death and of the individual in the attainment of sustenance. Spontaneous generation or the idea that life routinely arises from non-life was a common sense explanation of the miracle of life.

It had the advantage of simplicity, ease of understanding and didn't require any waste-of-time thinking. As science and the scientific method grew with the slow accumulation of knowledge, observant individuals began to consider the origin of life more deeply. Simple observations convinced many people that all the larger animals and plants produced life from previous life. Despite this, the mass of humans clung to the comfortable idea of spontaneous generation further, religions saw it as a convenient way to demonstrate the hand of god operating continuously in the world.

Some individuals, such as j.b. Van helment even described how one could make mice from grain, a jar and dirty rags by putting them together in a dark place for a few weeks and soon mice would appear in the jar.

Other, more perceptive individuals, like f. Redi tested the common idea that maggots arise via spontaneous generation on rotting meat. He placed a piece of meat in three jars, one he left open, one he corked tightly and the third he covered with a fine mesh gauze. Maggots only appeared in the open container, no matter how long he left the jars. Redi's experiment was important because of its eloquent simplicity. Anyone could repeat it and obtain the same clear results. Nevertheless, many people clung fiercely to the idea of spontaneous generation, while others designed experiments to test it.

In every case the results of the majority of these experiments indicated that spontaneous generation did not occur. The intellectual ferment this controversy stirred up gradually evolved into the scientific method as the various antagonists questioned each other's assumptions and, more importantly, their experimental design.

These arguments forced the designing of better experiments and eventually persuaded all but the most recalcitrant believers to discard spontaneous generation as an explanation for all higher life forms. Then, in the 1800's the refinement of the microscope, through which people could see tiny life forms that they assumed were simple, gave spontaneous generation proponents new life.

Again, flawed common sense led reasonable people astray. As the existence of microscopic life was accepted, the assumption was that such life must be simple compared to higher, more complex life.

The reasoning that followed this erroneous assumption was that since the microbes were small they must be simple & it followed that they were formed by spontaneous generation, hence god was still at work creating micro-life. Small is not simple!

The battle over spontaneous generation raged anew both from the pulpit and the lab. A number of scientists performed elementary experiments in which they treated soups and broth's, which left unheated would team with microbes after a few days, with heat to destroy any life present in them and asked the question: "Would new life arise in these sterile soups"? Spallanzani boiled "Soup" in glass containers and melted the glass closed.

The observation that nothing subsequently grew in this "heated" soup suggested that spontaneous generation didn't work. His detractors, rightly criticized his experiments, proposing that since air is necessary for life and since he had sealed the flask to air, obviously no life could develop. Others boiled soups and microbes grew, thus apparently supporting spontaneous generation. But again the preponderance of data suggested that spontaneous generation did not even apply to the "simple" microbes.

In 1859 one of the fathers of modern microbiology. Pasteur decided to settle the question of spontaneous generation once and for all. A genius at devising definitive experiments, pasteur first drew the necks of glass flasks out so that they remained open to the air, but were bent so that air could only enter by a curved path He then added broth and boiled it to destroy contaminating microbes. These flasks were then incubated and observed for months. He reasoned that the microbes in the air that could contaminate the sterile broth would be trapped on the sides of the thin glass necks before they reached the sterile broth. If spontaneous generation didn't occur no growth should take place.

This is exactly what happened, the flasks remained sterile indefinitely, until pasteur tipped the sterile broth up into the curved neck where he predicted the airborne organisms would have settled. After doing this the broths always grew microbes. These experiments ended the spontaneous generation controversy because these experiment was so elegant and simple and the results so clear, that anyone could repeat them.

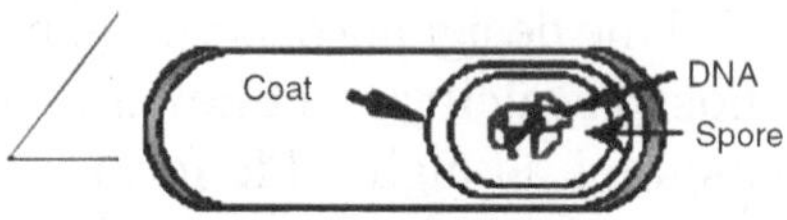

Fig. Spore Structure

Later, an earlier problem, in which occasional heated-broths did not remain sterile, was explained with the discovery of the heat resistant bacterial some of which could can survive several hours of boiling without being killed.

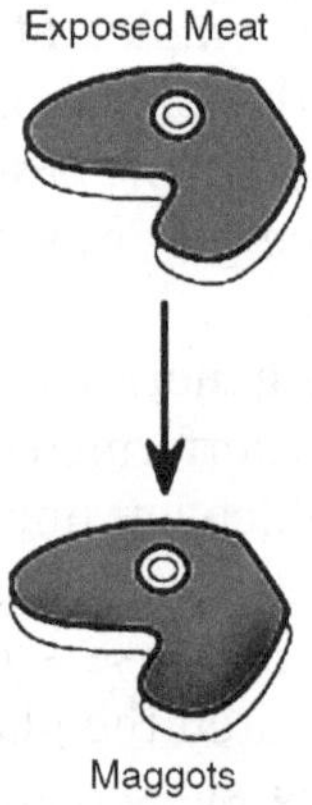

Fig. The Redi Experiment

Pasteur discovered many of the basic principles of microbiology and, along with R. Koch, laid the foundation for the science of microbiology. In 1857 Napoleon III was having trouble with his sailors mutinying because their wine was spoiling after only a few weeks at sea. Naturally Napoleon was distraught because his hopes for world conquest were being scuttled (pardon the pun) over a little spoiled wine, so he begged Pasteur for help.

Pasteur, armed with his trusty microscope, accepted the challenge and soon recognized that by looking at the spoiled wines he could distinguish between the contaminants that caused the spoilage and even predict the taste of the wine solely from his microscopic observations. He then reasoned that if one were to heat the wine to a point where its flavour was unaffected, but the harmful microbes were killed it wouldn't spoil.

As we are aware this process, today known as pasteurization, worked exactly the way he predicted and is the foundation of the modern treatment of bottled liquids to prevent their. It is important to realise that pasteurization is not the same as sterilization. Pasteurization only kills organisms that may spoil the product, but it allows many microbes to survive, whereas sterilization kills all the living organisms in the treated material.

Pasteur also realized that the yeast that was present in all the produced the alcohol in wine. When he announced this, a number of famous scientists were enraged, because the current theory of wine production was that wine formation was the result of spontaneous chemical changes that occurred in the grape juice. Pasteur was attacked furiously at scientific meetings, to the point where certain scientists did humorous skits about Pasteur and his tiny little yeast "stills" turning out alcohol. Pasteur had the last laugh however as people all over the world soon realized that if he was right they could control the quality of wine by controlling the yeast that made it. In a short period many others verified his observations and the opposition sank without a sound.

MICROBES

A microbe is any living thing that spends its life at a size visible sometimes only with a microscope. It is too tiny to be seen with the naked eye. Microbes are the oldest form of life on Earth. Some types have existed for billions of years. They may live as individuals or cluster together in communities. Microbes live in the water we drink, the food we eat and the air we breathe.

Right now, billions of microbes are swimming in our belly and mouth and crawling on our skin! Don't worry, over 95% of microbes are good for you. Microbes include bacteria, viruses, fungi, algae and protozoa. These single-cell organisms are invisible to the eye, but they can be seen with microscopes.

Fig. Microbe

CLASSIFICATION OF MICROBES

The term microbe is short for microorganism, which means small organism. To help people understand the different types of microbes, they are grouped or classified in various ways. Microbes are very diverse and represent all the great kingdoms of life. In fact, in terms of numbers, most of the diversity of life on Earth is represented by microbes.

Here is an outline of the major groups of microorganisms:

- Viruses
- Bacteria
- Algae
- Fungi
- Protozoa

Viruses

A virus is too small to be seen without a microscope. A virus is basically a tiny bundle of genetic material carried in a shell called the viral coat. Some viruses have an additional layer around this coat called an envelope. That's basically all there is to viruses.

There are thousands of different viruses that come in many shapes. Many are multi-sided or polyhedral. If you've ever looked closely at a cut gem, like the diamond in an engagement ring, you've seen an example of a polyhedral shape.

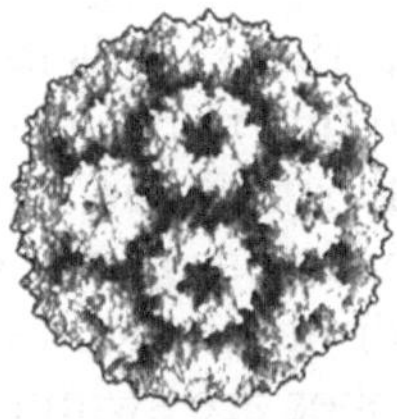

Fig. Viruses

Unlike the diamond in a ring, however, a virus does not taper to a point, but is shaped the same all around. Other viruses are shaped like spiky ovals or bricks with rounded corners. Some are like skinny sticks while others look like pieces of looped string. Some are more complex and shaped like little spaceship landing pods.

Viruses are found on or in just about every material and environment on Earth from soil to water to air. They're basically found anywhere there are cells to infect. Viruses can infect every living thing. However, viruses tend to be somewhat picky about what type of cells they infect. Plant viruses are not equipped to infect animal cells, for example, though a certain plant virus could infect a number of related plants. Sometimes, a virus may infect one animal and do no harm, but cause a great deal of damage when it gets into a different but closely related animal.

Viruses exist to reproduce only. To do that, they have to take over suitable host cells. Upon landing on a suitable host cell, a virus gets its genes inside the cell either by tricking the host cell to pull it inside, or by connecting its viral coat with the host cell wall or membrane and releasing its genes inside, or by injecting their genes into the host cell's DNA. The viral genes are then copied many times, using the process the host cell would normally use to reproduce its own DNA. The new viral genes then come together and assemble into whole new viruses. The new viruses are either released from the host cell without destroying the cell or eventually build up to a large enough number that they burst the host cell.

Bacteria

Bacteria consist of only one cell, but they're a very complex group of living things. Some bacteria can live in temperatures above the boiling point and in cold below the freezing point.

There are thousands of species of bacteria, but all of them are basically one of three different shapes. Some are rod- or stick-shaped; others are shaped like little balls. Others still are helical or spiral in shape. Some bacteria cells exist as individuals while others cluster together to form pairs, chains, squares or other groupings. Some bacteria can make their own food from sunlight, just like plants. Also like plants, they give off oxygen. Other bacteria absorb food from the material they live on or in. Some of these bacteria can live off iron or sulfur! The bacteria that live in our stomach absorb nutrients from the digested food you've eaten.

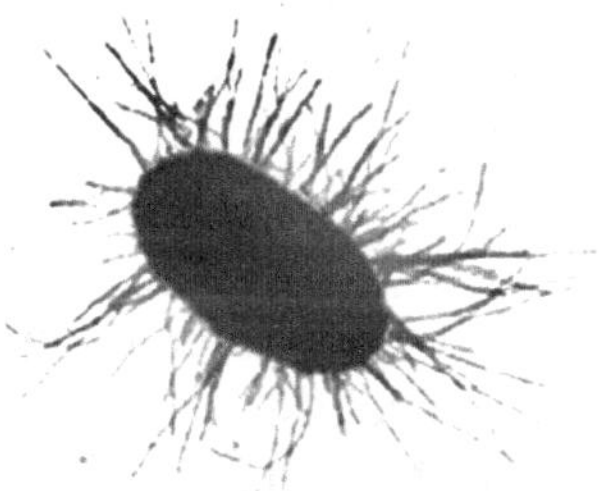

Fig. Bacteria

Some bacteria move about their environment by means of long, whip-like structures called flagella. They rotate their flagella like tiny outboard motors to propel themselves through liquid environments. They may also reverse the direction in which their flagella rotate so that they tumble about in one place.

Other bacteria secrete a slime layer and ooze over surfaces like slugs. Others stay almost in the same spot. Bacteria live on or in just about every material and environment on Earth from soil to water to air and from our body to the Arctic ice to the Sahara deserts. Each square centimeter of our skin averages about 100,000 bacteria. A single teaspoon of soil contains more than a billion (1,000,000,000) bacteria.

Algae

Algae are found in fresh and salt water around the world. They can also grow on rocks and trees and in soil when enough water is available.

Most algae are able to make energy from sunlight, like plants do. They produce a large amount of the oxygen we breathe. However, at some stages of their lives, some algae get their nutrients from other living things.

Diatoms are one kind of algae. They have hard shells made out of glass. When they die, these shells sink to the bottom of their watery environments. We mine deposits of these glass shells that formed hundreds of thousands of years ago to make abrasives, shiny road paint and grit in toothpaste. Diatoms come in all sorts of shapes—some are round and others are oval. Some look like leaves and others like fat commas.

Fig. Algae

Because algae make so much oxygen, these microbes are very helpful. But sometimes certain kinds of algae can also grow in such large numbers that when they suddenly die off en masse, the breaking down of their cells by bacteria destroys the amount of dissolved oxygen in the water, hurting the animals and plants that live there.

Fungi

Fungi are organisms that scientists once confused with plants. However, scientists have found that, at the cell level, the fungi are more like animals than they are like plants. For one thing, fungi cannot synthesize their own food like plants do, but instead they eat other organisms as do animals. Fungi come in a variety of shapes and sizes and different types. They can range from single cells to enormous chains of cells that can stretch for miles.

Fungi include single-celled living things that exist individually, such as yeast and multicellular clusters, such as molds or mushrooms. Yeast cells look round or oval under a microscope. They're too small to see as individuals, but we can see large clusters of them as a white powdery coating on fruits and leaves.

Molds are described as filament-like because they form long filament-like, or thread-like, strands of cells called hyphae. These hyphae are what give mold colonies their fuzzy appearance. They also form the fleshy body, or mushroom, that some species grow. It may seem strange to think of something as big as a mushroom as a microbe. But the cells of the hyphae making up that mushroom are connected in a closer way than the cells of other multicellular living things, like we and me, are.

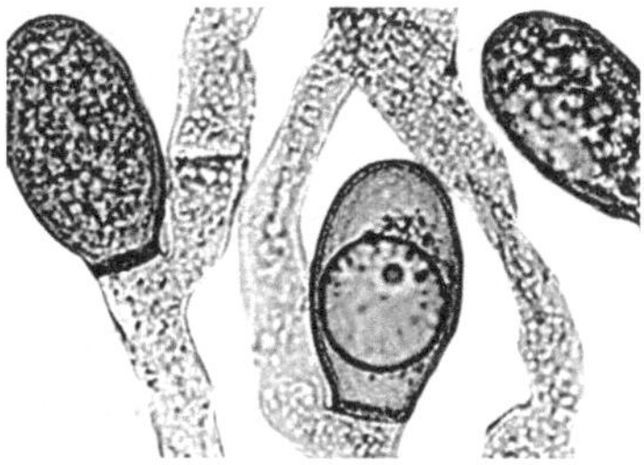

Fig. Fungi

Fungi usually grow best in environments that are slightly acidic. They can grow on substances with very low moisture. Fungi live in the soil and on our body, in our house and on plants and animals, in freshwater and seawater. A single teaspoon of topsoil contains about 120,000 fungi.

Fungi are basically stationary. But they can spread either by forming reproductive spores that are carried on wind and rain or by growing and extending their hyphae. Hyphae grow as new cells form at the tips, creating even longer chains of cells. Fungi absorb nutrients from living or dead organic matter that they grow on. They absorb simple, easily dissolved nutrients, such as sugars, through their cell walls. They give off special digestive enzymes to break down complex nutrients into simpler forms that they can absorb.

Some fungi are quite useful to us. We've used several kinds to make antibiotics to fight bacterial infections. These antibiotics are based on natural compounds the fungi produce to compete against bacteria for nutrients and space. We use baker's yeast, to make bread rise and to brew beer. Fungi break down dead plants and animals and keep the world tidier. We're exploring ways to use natural fungal enemies of insect pests to get rid of these bugs.

There are some dangerous fungi that cause diseases in plants, animals and people. Fungi ruin about a quarter to half of harvested fruits and vegetables each year.

MICROBES ARE HELPFUL OR NOT

Microbes are much more our friends than our enemies. Although some microbes cause health problems such as strep throat, chickenpox and the common cold, most microbes make our lives better such as: Bacillus thuringiensis - a common soil bacterium that is a natural pest-killer in gardens and on crops. Arbuscular mycorrhizas - fungus living in the soil that helps crops take up nutrients from the soil. Accharomyces cerevisiae - baker's yeast that makes bread rise.

Escherichia coli - one of many kinds of microbes that live in our digestive system to help we digest our food every day. Streptomyces - bacteria in soil that makes an antibiotic used to treat infections. Pseudomonas putida - one of many microbes that clean wastes from sewage water at water treatment plants. Lactobacillus acidophilus - one of the bacteria that turn milk into yogurt.

There are many other important jobs microbes do. They are used to make medicine. They break down the oil from oil spills. They make about half of the oxygen we breathe. They are the foundation of the food chain that feeds all living things on earth. We've been using microbes for thousands of years to make products we need and enjoy. For example, we can thank fungi for the cheese on our cheeseburger and yeast for our bun. Cheese and bread are two microbe-made foods people have been enjoying since time began.

In pollution control, researchers are using bacteria that eat methane gas to clean up hazardous waste dumps and landfills. These methane-eating bacteria make an enzyme that can break down more than 250 pollutants into harmless cells.

By piping methane into the soil, researchers can increase growth of the bacteria that normally live in the polluted soil. More bacteria means faster pollution break up.

Also, bacteria is being used as one of the tools to clean up oil spills. These bacteria eat the oil, turning it into carbon dioxide and other harmless by-products. Fungi and bacteria produce antibiotics such as penicillin and tetracycline.

These are medicines we use to fight off harmful bacteria that cause sore throats, ear infections, diarrhea and other discomforts. Scientists have changed the genetic material of bacteria and yeasts to turn them into medicine. They inject genes for medicines they want to make into the microbe cells, as if adding new building information to the microbe's cell DNA. The scientists then grow the microbes in huge containers called fermenters where they reproduce into billions, all making new medicines.

Microbes and the Days of Creation

The world of germs and microbes has received much attention in recent years—and for good reason. We frequently hear the term microbe associated with organisms such as Escherichia coli, Salmonella, anthrax bacteria, antibiotic-resistant tuberculosis, MRSA, HIV, malaria, Stachybotrys and other microscopic creatures. But where do microbes fit into the creation account? Were they created along with the rest of the plants and animals in the first week of creation, or were they created later, after the Fall?

Are microbes a result of the Curse? These and other questions are some that a group of professional creation microbiologists have been asking and their answers may surprise us. Ongoing research based on the creation paradigm appears to provide some answers to these puzzling questions.

Although we cannot be dogmatic about the details of microbe origin during Creation Week, it is believed that a reasonable extrapolation from biological data and Scripture can be made about the nature of microbes in a fully mature creation. Past creation scientists such as Leeuwenhoek, Pasteur and Lister, were blessed by God as He revealed to them critical insight into His creation. So, where do these microbes fit into the very good days of creation? Before answering this question, three terms, microbe, germ and symbiosis, need to be defined. These are relatively new or "modern" terms. First, the Bible does not use these specific terms. These terms were not commonly used until the end of the nineteenth century.

The term microbe was first used in 1878 to describe "extremely minute living beings." Before 1878, scientists including Louis Pasteur, used a variety of terms rather loosely to label the very small organisms that had interested them. It was not clear whether microbes belonged to the animal or plant kingdom, or to a completely different one. The term microbe was given by Charles E. Sedillot to describe bacteria Later, it would also be used of eukaryotic cells, including algae, fungi, protozoans and slime molds. Some people refer to viruses as microbes, but others do not because viruses are not cells. Viruses have nonliving, as well as living characteristics.

Today, the term germ refers to disease-causing microbes, or pathogens. All germs would have originated after the Fall The Edenic Curse would have profoundly influenced all creation, including viruses, bacteria, fungi and protozoans that would later become pathogens or parasites. The origin of infectious disease is complex and multifaceted. This topic is further explored in the book.

The Genesis of Germs It provides some understanding into the origin of infectious diseases. From a biblical worldview, infectious diseases and pathogenesis are a secondary state in nature. It is not the way the Creator intended for man and nature. Most microbes are beneficial to man and nature. Only about 5–10% of all bacteria are pathogenic. Many microbes live in a mutualistic relationship with plants, animals, humans and other microbes. Mutualism is a type of symbiosis. The term symbiosis is used to describe an intimate association between organisms of different species. Some symbiotic relationships between microbes and plants, animals and humans are essential for life on earth.

For this reason most microbiologists maintain that bacteria, fungi, protists and other microbes have been maligned in the news media. Without our intestinal flora, we would not digest food nor acquire vitamins and minerals very efficiently. Without fungi, bacteria, algae and protozoans, life on earth could not last. This is because microbes provide essential "services" (e.g., nitrogen fixation, nitrification and denitrification) in nutrient cycles.

Viruses Fit into Creation

The determination of virus origin is uncertain. It may be that viruses have multiple origins. Some may be degenerate parts from cells after the Curse; still others may have their origin during the days of creation. Today, we think of viruses (Latin for "poison") only in the context of disease. However, some viruses (or at least virus-like genes) are involved in a positive function in nature.

Some groups of viruses, like bacteriophages, play a positive role in controlling bacteria in ecosystems and may play a role in diversity. Another group of viruses play a role in turning off the immune system during pregnancy in mammals and humans. This is a group referred to as endogenous retroviruses (ERVs). ERVs are among a kind of repetitious genetic elements called "retrotransposons". Research has shown that the ERV design prohibits the mother's immune system from damaging the child's body.

These retroviruses cannot fully replicate, only expressed in local immune cells (such as macrophages) of the placenta, thereby preventing them from initiating a full-blown immune response. Thus, the mother's immune system remains competent to respond to other infections but is specifically prevented from mounting an immune response to the developing embryo.

So in creation, the selective ability to turn off the immune system for protection would be a "good" design. Other ERVs also play a positive role in animal and human reproduction. However, since the corruption of creation, the corrupted retrovirus, HIV and various leukemia viruses turn off the entire immune system, leaving the body open to devastating infections.

These examples may provide clues to the origin of viruses and how some may have been created during Creation Week by design and how some have been corrupted as a result of the Fall.

Aseptic Technique

Trying to study this mixed population is often difficult and in the tradition of the scientific method; researchers dissect a system and study each piece in isolation. For microorganisms this means separating the organisms and getting them into pure culture. A pure culture is defined as a growth of microorganisms (a culture) that contains one cell type. It is essential in microbiology to be able to obtain and preserve pure cultures.

Over 100 years ago, Robert Koch devised methods to achieve this goal and the methods he developed are essentially still used today. The goals of aseptic technique are two-fold. The first objective is to obtain pure cultures and secondly to prevent cross-contamination. Microorganisms in culture must not escape into the environment and microbes in the environment must not get into the cultures we are studying. It is essential that aseptic technique be understood and practiced correctly.

Contaminated cultures are worthless for diagnosis or for doing research on, because it is unclear what microbe is performing any action that is being observed. Aseptic methods commonly used are flame sterilization, tube transfer, streak plates, spread plates and pour plates. Flame sterilization is an easy method to insure sterile transfer of a culture from a source to a growth medium.

Tube transfer is useful for moving inocula from one tube to another. Mechanical dilution by making streak plates is the preferred method for obtaining a pure culture of a microorganism. Finally, spread plates and pour plates are common methods for enumerating microorganisms and are sometimes useful for obtaining isolated colonies.

Flame Sterilization and Tube Transfer

Flame sterilization is a very quick simple method of killing microorganisms on an inoculating loop or needle. The loop or needle is held inside a flame for a few seconds to bring it to redness and then cooled. Once cool, the loop or needle can be used for various culture manipulations. Make sure that the area that contacts the culture is flamed to redness. Also, be patient and let the loop cool down, this usually takes about 15-30 seconds. Learning this technique is essential to everything else we do in microbiology.

Transfer of culture from agar plates to tubes, or from tube to tube, is a common, simple procedure. It is important to perform these transfers in a consistent and rapid manner. The following protocols have been found effective. To transfer a culture from an agar plate to a broth or agar slant:

- Place the Bunsen burner in front of we and assemble all necessary equipment with in arms reach. Position everything so that as suggested, not burn ourself while trying to inoculate our tubes.
- Label the tube of broth or agar to be inoculated with identifying marks. The culture, the date and our initials for example. Place it in a rack in front of we.
- Holding the inoculating loop handle, flame the entire wire to redness.
- When the wire cools (about 15-30 seconds) remove the lid of the plate with our other hand and obtain an inoculum by removing a small portion of the surface growth on the agar plate. In most cases as suggested, be picking an isolated colony. Choose a well isolated one. Do not dig into the agar. Replace the lid of the plate immediately.
- Hold the tube to be inoculated with the free hand. Remove the cotton plug or cap of the tube with the little finger of the hand holding the needle holder. If a cotton-plugged tube is used, the mouth of the tube should be passed briefly through the flame to singe off dust and lint particles. (Dust or lint may fall into the tube and contaminate the medium.)
- Introduce the inoculum into the tube. When inoculating a tube of broth, rub the wire against the glass just above the fluid level and then tip the tube slightly to wash the inoculum into the broth. The wire should not be rattled against the sides of the tube to shake an inoculum into the broth; this is unnecessary and may create a

dangerous aerosol. If the transfer is made to an agar slant, a single mid-line stroke over its surface is made with the wire or loop.

- Replace the cap or plug.
- Flame the inoculating wire again to redness, slowly to avoid spattering. Put the loop holder down after the wire cools.

Making a Medium

Making medium is as simple as cooking and a crude medium can be made in almost any kitchen with a few utensils and a source of heat. The production of a chicken broth medium that will grow many common microorganisms. The medium will be sterilized by tyndaliza-tion. Simply boiling a medium once, while it will kill most vegetative cells, does not kill endospores. However, heating encourages the endospores to germinate and a second boiling kills these microbes. The boiling process is repeated a third time to ensure that all spores have germinated and been killed.

- Add 250 ml (about 1 cup) of water into a glass container or some other vessel that can stand boiling water. The container should be something we can cover. Glass bottles that can stand boiling or canning jars work well.
- To this add 15 grams, about 1 tablespoon of instant chicken broth crystals. and stir until dissolved. Cover loosely so that steam can escape, but dust and dirt cannot enter.
- Microwave the broth on high for 3 minutes or until it just begins to boil.
- Let the broth cool and sit for at least 2 hours up to overnight at room temperature. Make sure that the medium vessel is covered to prevent contamination from the air. At this point the vegetative cells are dead and most endospores will germinate.
- Repeat steps 3 and 4 a second time. This will kill the endospores that germinated on day 1.
- Repeat steps 3 and 4 a third time. This will eliminate the remaining endospores, making the medium sterile.
- Place the medium in a warm place overnight. If our tyndalization was done correctly, our medium should remain clear and free of microbes.
- While we are waiting for our medium, search for a sample we would like to inoculate into it. Almost any sample should work, but natural samples are a good idea for this experiment. Samples coming from food or from our body may contain pathogens. Something we probably want to avoid.

- Check our medium. If it is still clear and sterile, add a pinch of our collected sample to the medium. (The exact amount really does not matter). Again place the medium in a warm place and check it periodically. After a few days, we should start to see the medium become more turbid.
- If we have the equipment, examine a sample of the medium under a microscope. What shapes of microbes are present? Is there more than one species?
- When finished, carefully dispose of the borth. It is basically a spoiled food, but the properties of the spoilage microbe are unknown and should be treated with care.

STREAK PLATES

The streak plate method is a rapid and simple technique of mechanically diluting a relatively large concentration of microorganisms to a small, scattered population of cells. The goal is to obtain isolated colonies on a large part of the agar surface, so that desired species can then be brought into pure culture.

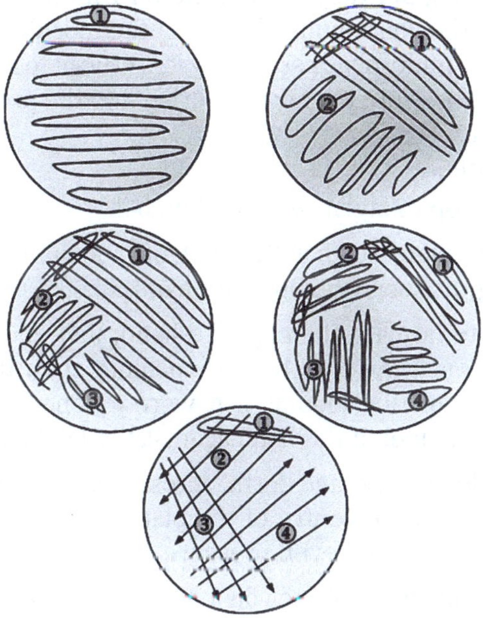

Fig. Streaking Patterns

Proper streaking of plates is an indispensable tool in microbiology. In most cases a closed inoculating loop is used for streaking plates. The wire loop should not be badly oxidized or pitted or it will fail to dilute the inoculum and will scratch the surface of the agar. Streak plates can be made from a broth culture, an agar slant or from an agar plate. It is sometimes convenient to suspend a bit of growth from a solid surface in sterile saline and use this as a source of inoculum. Resuspension of colonies or cultures grown on solid surfaces dilutes the culture and makes streak plating easier.

A loopful of inoculum is transferred from the source and put on the agar surface. When using a large inoculum (a turbid culture or growth from a solid surface), a small spot is spread during the initial transfer. If the inoculum is from a lightly turbid suspension, the first phase of the streaking pattern is begun. The three-phase streaking pattern is recommended for beginners because it is most likely to give satisfactory results with suspensions having a wide range of microbial density. There are a number of different methods for mechanically diluting microbes on a streak plate. The most common method is spreading microbes across a plate. As the concentration of microbes increases so do the number of phases. Irrespective of the number of phases, loop is flamed between each one. The fifth plate shows an alternative method, where the streaks are not continuous, but are a series of parallel lines. Foobar.

Choosing a streaking pattern is a matter of individual preference and depends upon the number of microorganisms in the sample. Demonstrates the most common patterns, but they are not the only methods. The object of any streaking pattern is the continuous dilution of the inoculum to give many well isolated colonies. For multi-phase streaking it is crucial to flame the loop before starting the next phase. Note the slight overlap into the previous phase to pick up a small inoculum.

To streak a plate...

- Flame the loop to sterilize it and let cool.
- Position the plate so that the spot of inoculum is nearest the hand not holding the loop (the opposite hand).
- Lift the plate lid with the opposite hand; just enough to get the loop inside and touch the loop to the inoculum spot. It is often helpful to treat the inoculating loop as if it were a pencil - steadying the loop by resting the heel of the hand against the lab bench.
- Move the loop back and forth across the spot and then gradually continue toward the centre of the plate as we sweep back and forth. Use a very gentle and even pressure.
- When creating each phase, do not worry about keeping each pass across the plate separate from previous ones.
- When about 30% of the plate has been covered by the first streaking phase, remove the loop and flame sterilize it.
- Repeat the above procedure for the second phase, but this time pick up some inoculum by crossing into the first phase 2-3 times and then not passing into it again.

- Repeat as necessary for the third and fourth phases. After streaking the plate, flame sterilize the loop before setting it down.

Spread Plates and Dilution Plating

An absolute requirement for a microbiologist is to be able to determine the concentration of microorganisms in a given sample. Various particle-counting devices, spectrophotometric methods and microscopic techniques have been used to count cells. However, one drawback to these methods is that they count dead as well as living cells.

The most common method of enumerating viable cells is the plate-count method. Diluting microorganisms and placing them into petri plates (or plates) for incubation is another essential technique for working with microorganisms. This method suffers from some problems. First, only those organisms which can grow on the medium and at the temperature and atmospheric conditions of incubation, will divide and develop into colonies. Second, each colony may not represent the progeny from one cell, as two or more cells (those in clusters, chains or otherwise close to one another) can give rise to one colony.

For these reasons the counts obtained from the plate-count method are given as the number of colony-forming units (CFU's) per ml (or gram) rather than the number of cells per ml (or gram). Despite these drawbacks, the plate-count method is a powerful means by which concentrations of viable organisms may be estimated.

Also, if it is desirable to count a specific subgroup of microorganisms in a sample, selective media or special incubation conditions can often be used to encourage the growth of only this class of organism.As microbial quantitation involves the use of pipettes (or micropipettes) in preparing dilutions and inoculating plates, the beginning student in microbiology must become familiar with their use.

Due to the possibility of ingesting pathogens and toxic liquids, mouth-pipetting is forbidden in the laboratory! Pipettes are filled and subsequently emptied by the use of propipettes or other pipette bulb. Pay close attention to the demonstration of their use. Important Safety Consideration: When fitting the pipette and pipette bulb together, use very gentle pressure!! Do not jam these items together! (Force is usually not the answer, a good general rule to live by in this lab and in life for that matter.) The glass pipette will probably break and possibly cause severe injury. Handle the pipette only at the top inch or so. For volumes of 5 ml of less, micropipettes are often the tool of choice.

Instruments are available that are capable of dispensing 5 ml all the way to less than 1 μl (one one-millionth of a liter). Micropipettes have made it possible to miniaturize many experiments and greatly decrease the cost of running them. They are also easy to use and can dispense volumes quickly, increasing the number of experiments that can be performed in a set amount of time. Micropipettes are the tool of choice for small volumes.

MICROBIAL CULTIVATION

When microorganisms are cultivated in the laboratory, a growth environment called a medium is used. The medium may be purely chemical (a chemically defined medium), or it may contain organic materials, or it may consist of living organisms such as fertilized eggs.Microorganisms growing in or on such a medium form a culture. A culture is considered a pure culture if only one type of organism is present and a mixed culture if populations of different organisms are present. When first used, the culture medium should be sterile, meaning that no form of life is present before inoculation with the microorganism.

General microbial media. For the cultivation of bacteria, a commonly used medium is nutrient broth, a liquid containing proteins, salts and growth enhancers that will support many bacteria. To solidify the medium, an agent such as agar is added. Agar is a polysaccharide that adds no nutrients to a medium, but merely solidifies it. The medium that results is nutrient agar. Many media for microorganisms are complex, reflecting the growth requirements of the microorganisms.

For instance, most fungi require extra carbohydrate and an acidic environment for optimal growth. The medium employed for these organisms is potato dextrose agar, also known as Sabouraud dextrose agar. For protozoa, liquid media are generally required and for rickettsiae and viruses, living tissue cells must be provided for best cultivation.

For anaerobic microorganisms, the atmosphere must be oxygen free. To eliminate the oxygen, the culture media can be placed within containers where carbon dioxide and hydrogen gas are generated and oxygen is removed from the atmosphere. Commercially available products achieve these conditions. Anaerobic chambers can also be used within closed compartments and technicians can manipulate culture media within these chambers. To encourage carbon dioxide formation, a candle can be burned to use up oxygen and replace it with carbon dioxide.

Special microbial media. Certain microorganisms are cultivated in selective media. These media retard the growth of unwanted organisms while encouraging the growth of the organisms desired. For example, mannitol salt agar is selective for staphylococci because most other bacteria cannot grow in its high-salt environment. Another selective medium is brilliant green agar, a medium that inhibits Gram-positive bacteria while permitting Gram-negative organisms such as *Salmonella* species to grow.

Still other culture media are differential media. These media provide environments in which different bacteria can be distinguished from one another. For instance, violet red bile agar is used to distinguish coliform bacteria such as *Escherichia coli* from noncoliform organisms. The coliform bacteria appear as bright pink colonies in this media, while noncoliforms appear a light pink or clear. Certain media are both selective and differential. For instance, MacConkey agar differentiates lactose-fermenting bacteria from nonlactose-fermenting bacteria while inhibiting the growth of Gram-positive bacteria. Since lactose-fermenting bacteria are often involved in water pollution, they can be distinguished by adding samples of water to MacConkey agar and waiting for growth to appear.

In some cases, it is necessary to formulate an enriched medium. Such a medium provides specific nutrients that encourage selected species of microorganisms to flourish in a mixed sample. When attempting to isolate *Salmonella* species from fecal samples, for instance, it is helpful to place a sample of the material in an enriched medium to encourage *Salmonella* species to multiply before the isolation techniques begin.

In order to work with microorganisms in the laboratory, it is desirable to obtain them in pure cultures. Pure cultures of bacteria can be obtained by spreading bacteria out and permitting the individual cells to form masses of growth called colonies.

One can then pick a sample from the colony and be assured that it contains only one kind of bacteria. Cultivating these bacteria on a separate medium will yield a pure culture.

To preserve microbial cultures, they may be placed in the refrigerator to slow down the metabolism taking place. Two other methods are deep-freezing and freeze-drying.

For deep-freezing, the microorganisms are placed in a liquid and frozen quickly at temperatures below –50°C. Freeze-drying (lyophilization) is performed in an apparatus that uses a vacuum to draw water off after the microbial suspension has been frozen. The culture resembles a powder and the microorganisms can be preserved for long periods in this condition.

STERILIZATION

Wet Heat

Wet heat is the most dependable procedure for the destruction of all forms of microbial life. Steam sterilization generally denotes heating in an autoclave employing saturated steam under a pressure of approximately 15 psi to achieve a chamber temperature of at least 121°C (250°F). The critical factors in insuring the reliability of this sterilization method is:

- Proper temperature and time; and
- The complete replacement of the air with steam (i.e. no entrapment of air).

Some autoclaves utilize a steam activated exhaust valve that remains open during the replacement of air by live steam until the steam triggers the valve to close. Others utilize a pre-cycle vacuum to remove air prior to steam introduction. Physical controls such as pressure gauges and thermometers are widely used but are considered secondary methods of insuring sterilization.

The use of appropriate biological indicators at locations throughout the autoclave is considered the best indicator of sterilization. The biological indicator most widely used for wet heat sterilization is Bacillus stearothermophilus spores.

Dry Heat

Dry heat is less efficient than wet heat sterilization and requires longer times and/or higher temperatures. The specific times and temperatures must be determined for each type of material being sterilized. Generous safety factors are usually added to allow for the variables that can influence the efficiency of this method of sterilization.

The moisture of the sterilization environment as well as the moisture history of organisms prior to heat exposure appear to affect the efficiency of dry heat sterilization. Higher temperatures and shorter times may be used for heat resistant materials. The heat transfer properties and the spatial relation or arrangement of articles in the load are critical in insuring effective sterilization.

The advantage of wet heat is a better heat transfer to and into the cell resulting in overall shorter exposure time and lower temperature. Steam sterilization uses pressurized steam at 121-132° C (250-270° F) for 30 or 40 minutes. This type of heat kills all microbial cells including spores, which are normally heat resistant. In order to accomplish the same effect with dry heat in an oven, the temperature needs to be increased to 160-170° C (320-338° F) for periods of 2 to 4 hours.

General Procedures

All materials, equipment, apparatus contaminated with or containing potentially hazardous organisms should be steam sterilized before being washed and stored, or discarded. Autoclaving is the preferred method. Each individual working with biohazardous materials is responsible for sterilization of materials before disposal.

Biohazardous materials should not be placed in autoclaves overnight in anticipation of autoclaving the next day. To minimize hazard to firemen or disaster crews, all biohazardous materials must be placed in an appropriately marked refrigerator or incubator, sterilized, or otherwise confined at the close of each work day.

All autoclaves must be certified for operating efficiency by the periodic use of biological indicator controls and records maintained for three years. Special precautions should be taken to prevent accidental removal of material from an autoclave before it has been sterilized or the simultaneous opening of both doors on a double door autoclave. Dry hypochlorites, or any other strong oxidizing material, must not be autoclaved with organic materials such as a paper, cloth, or oil:

Oxidizer + Organic material + Heat = Possible Explosion.

All floors, laboratory benches and other surfaces in buildings where biohazardous materials are handled should be disinfected as often as deemed necessary by the supervisor.

The surroundings should be disinfected after completion of operations involving plating, pipetting, centrifuging and similar procedures with biohazardous materials.

It is the responsibility of the supervisor to determine that the disinfectant and the time and method of exposure is effective against the biological agent(s) used in the facility.

Floor drains should be flooded with water or disinfectant at least once each week in order to fill traps and thus prevent the back flow of sewer gases. Floor cleaning procedures which minimize the generation of aerosols should be used. Wet mopping or wet vacuum pick up is recommended.

Water used to mop floors should contain a disinfectant or disinfectant-detergent. (Dry mopping or dusting should be avoided). Where wet procedures are not practical, dry vacuum cleaning with a HEPA filter on the exhaust, sweeping compound used with push brooms, or dry dust mop heads treated to suppress aerosolization may be used. Stock solutions of suitable disinfectants will be maintained in each laboratory for disinfection purposes. NOTE: working dilutions of certain disinfectants have short shelf lives and must be prepared on a regular basis.

Sterilization Procedures

Supervisors are encouraged to review the type of materials being handled and to establish standard conditions for sterilization. Treatment conditions to achieve sterility will vary in relation to the volume of material treated, volume of the autoclave, the contamination level, the moisture content and other factors.

Steam Autoclave:

- *Laundry:* 250°F (121°C) for a minimum of 30 min.
- *Trash:* 250°F (121°C) for at least 45 minutes per bag. Size of the autoclave and of the bags greatly effect sterilization time. Large bags in a small autoclave may require 90 minutes or more.
- *Glassware:* 250°F (121°C) for a minimum of 25 min.
- *Liquids:* 250°F (121°C) for 25 minutes for each gallon.
- *Animals & bedding:* Steam autoclaving not recommended (sterilization time required would be at least 8 hours). Incineration in an approved facility is the recommended treatment of these wastes.

MICROSCOPES

Since microorganisms are invisible to the unaided eye, the essential tool in microbiology is the microscope. One of the first to use a microscope to observe microorganisms was Robert Hooke, the English biologist who observed algae and fungi in the 1660s. In the 1670s, Anton van Leeuwenhoek, a Dutch merchant, constructed a number of simple microscopes and observed details of numerous forms of protozoa, fungi and bacteria.

During the 1700s, microscopes were used to further elaborate on the microbial world and by the late 1800s, the sophisticated light microscopes had been developed. The electron microscope was developed in the 1940s, thus making the viruses and the smallest bacteria (for example, rickettsiae and chlamydiae) visible. Microscopes permit extremely small objects to be seen, objects measured in the metric system in micrometers and nanometers.

A micrometer (ìm) is equivalent to a millionth of a meter, while a nanometer (nm) is a billionth of a meter. Bacteria, fungi, protozoa and unicellular algae are normally measured in micrometers, while viruses are commonly measured in nanometers. A typical bacterium such as *Escherichia coli* measures about two micrometers in length and about one micrometer in width.

Optical Microscopy

Optical or light microscopy involves passing visible light transmitted through or reflected from the sample through a single or multiple lenses to allow a magnified view of the sample.

The resulting image can be detected directly by the eye, imaged on a photographic plate or captured digitally. The single lens with its attachments, or the system of lenses and imaging equipment, along with the appropriate lighting equipment, sample stage and support, makes up the basic light microscope.

Limitations of Optical Microscopy

Limitations of standard optical microscopy (bright field microscopy) lie in three areas;

- The technique can only image dark or strongly refracting objects effectively.
- Diffraction limits resolution to approximately 0.2 micrometre.
- Out of focus light from points outside the focal plane reduces image clarity.

Live cells in particular generally lack sufficient contrast to be studied successfully, internal structures of the cell are colourless and transparent. The most common way to increase contrast is to stain the different structures with selective dyes, but this involves killing and fixing the sample.

Staining may also introduce artifacts, apparent structural details that are caused by the processing of the specimen and are thus not a legitimate feature of the specimen.

These limitations have, to some extent, all been overcome by specific microscopy techniques which can non-invasively increase the contrast of the image. In general, these techniques make use of differences in the refractive index of cell structures. It is comparable to looking through a glass window: we (bright field microscopy) don't see the glass but merely the dirt on the glass.

There is however a difference as glass is a more dense material and this creates a difference in phase of the light passing through. The human eye is not sensitive to this difference in phase but clever optical solutions have been thought out to change this difference in phase into a difference in amplitude (light intensity).

OPTICAL MICROSCOPY TECHNIQUES

Bright Field Optical Microscopy

Bright field microscopy is the simplest of all the light microscopy techniques. Sample illumination is via transmitted white light, i.e. illuminated from below and observed from above. Limitations include low contrast of most biological samples and low apparent resolution due to the blur of out of focus material. The simplicity of the technique and the minimal sample preparation required are significant advantages.

Oblique Illumination and what it Means

The use of oblique (from the side) illumination gives the image a 3-dimensional appearance and can highlight otherwise invisible features. A more recent technique based on this method is Hoffmann's modulation contrast, a system found on inverted microscopes for use in cell culture. Oblique illumination suffers from the same limitations as bright field microscopy (low contrast of many biological samples; low apparent resolution due to out of focus objects), but may highlight otherwise invisible structures.

Dark field Optical Microscopy

Dark field microscopy is a technique for improving the contrast of unstained, transparent specimens. Darkfield illumination uses a carefully aligned light source to minimise the quantity of directly-transmitted (un-scattered) light entering the image plane, collecting only the light scattered by the sample.

Darkfield can dramatically improve image contrast—especially of transparent objects—while requiring little equipment setup or sample preparation. However, the technique does suffer from low light intensity in final image of many biological samples and continues to be affected by low apparent resolution.

Rheinberg illumination is a special variant of dark field illumination in which transparent, colored filters are inserted just before the condenser so that light rays at high aperture are differently colored than those at low aperture (i.e. the background to the specimen may be blue while the object appears self-luminous yellow). Other colour combinations are possible but their effectiveness is quite variable.

Phase Contrast Optical Microscopy

More sophisticated techniques will show differences in optical density in proportion. Phase contrast is a widely used technique that shows differences in refractive index as difference in contrast. It was developed by the Dutch physicist Frits Zernike in the 1930s The nucleus in a cell for example will show up darkly against the surrounding cytoplasm. Contrast is excellent; however it is not for use with thick objects. Frequently, a halo is formed even around small objects, which obscures detail.

The system consists of a circular annulus in the condenser which produces a cone of light. This cone is superimposed on a similar sized ring within the phase-objective. Every objective has a different size ring, so for every objective another condenser setting has to be chosen.

The ring in the objective has special optical properties: it first of all reduces the direct light in intensity, but more importantly, it creates an artificial phase difference of about a quarter wavelength. As the physical properties of this direct light have changed, interference with the diffracted light occurs, resulting in the phase contrast image.

Differential Interference Contrast Microscopy

Superior and much more expensive is the use of interference contrast. Differences in optical density will show up as differences in relief. A nucleus within a cell will actually show up as a globule in the most often used differential interference contrast system according to Georges Nomarski.

However, it has to be kept in mind that this is an optical effect and the relief does not necessarily resemble the true shape! Contrast is very good and the condenser aperture can be used fully open, thereby reducing the depth of field and maximizing resolution.

The system consists of a special prism in the condenser that splits light in an ordinary and an extraordinary beam. The spatial difference between the two beams is minimal (less than the maximum resolution of the objective). After passage through the specimen, the beams are reunited by a similar prism in the objective.

In a homogeneous specimen, there is no difference between the two beams and no contrast is being generated. However, near a refractive boundary (say a nucleus within the cytoplasm), the difference between the ordinary and the extraordinary beam will generate a relief in the image. Differential interference contrast requires a polarized light source to function; two polarizing filters have to be fitted in the light path, one below the condenser (the polarizer) and the other above the objective (the analyzer).

Note: In cases where the optical design of a microscope produces an appreciable lateral separation of the two beams we have the case of classical interference microscopy, which does not result in relief images, but can nevertheless be used for the quantitative determination of mass-thicknesses of microscopic objects.

FLUORESCENCE MICROSCOPY

When certain compounds are illuminated with high energy light, they then emit light of a different, lower frequency. This effect is known as fluorescence. Often specimens show their own characteristic autofluorescence image, based on their chemical makeup.

This method is of critical importance in the modern life sciences, as it can be extremely sensitive, allowing the detection of single molecules. Many different fluorescent dyes can be used to stain different structures or chemical compounds. One particularly powerful method is the combination of antibodies coupled to a fluorochrome as in immunostaining. Examples of commonly used fluorochromes are fluorescein or rhodamine. The antibodies can be made tailored specifically for a chemical compound.

For example, one strategy often in use is the artificial production of proteins, based on the genetic code (DNA). These proteins can then be used to immunize rabbits, which then form antibodies which bind to the protein. The antibodies are then coupled chemically to a fluorochrome and then used to trace the proteins in the cells under study.

Highly-efficient fluorescent proteins such as the green fluorescent protein (GFP) have been developed using the molecular biology technique of gene fusion, a process which links the expression of the fluorescent compound to that of the target protein. Piston DW, Patterson GH, Lippincott-Schwartz J, Claxton NS, Davidson MW Nikon MicroscopyU: Introduction to Fluorescent Proteins. Nikon MicroscopyU. Retrieved on 2007-08-22. This combined fluorescent protein is generally non-toxic to the organism and rarely interferes with the function of the protein under study. Genetically modified cells or organisms directly express the fluorescently-tagged proteins, which enables the study of the function of the original protein in vivo.

Since fluorescence emission differs in wavelength (colour) from the excitation light, a fluorescent image ideally only shows the structure of interest that was labelled with the fluorescent dye. This high specificity led to the widespread use of fluorescence light microscopy in biomedical research. Different fluorescent dyes can be used to stain different biological structures, which can then be detected simultaneously, while still being specific due to the individual colour of the dye. To block the excitation light from reaching the observer or the detector, filter sets of high quality are needed. These typically consist of an excitation filter selecting the range of excitation wavelengths, a dichroic mirror and an emission filter blocking the excitation light.

Most fluorescence microscopes are operated in the Epi-illumination mode to further decrease the amount of excitation light entering the detector.

Confocal Laser Scanning Microscopy

Generates the image by a completely different way than the normal visual bright field microscope. It gives slightly higher resolution, but most importantly it provides optical sectioning without disturbing out-of-focus light degrading the image. Therefore it provides sharper images of 3D objects. This is often used in conjunction with fluorescence microscopy.

Deconvolution Microscopy

Fluorescence microscopy is extremely powerful due to its ability to show specifically labelled structures within a complex environment but also because of its inherent ability to provide three dimensional information of biological structures. Unfortunately this information is blurred by the fact, that upon illumination all fluorescently labeled structures emit light no matter if they are in focus or not.

This means, that an image of a certain structure is always blurred by the contribution of light from structures which are out of focus. This phenomenon becomes apparent as a loss of contrast especially when using objectives with a high resolving power, typically oil immersion objectives with a high numerical aperture.

Fortunately though, this phenomenon is not caused by random processes such as light scattering but can be relatively well defined by the optical properties of the image formation in the microscope imaging system. If one considers a small fluorescent light source (essentially a bright spot), light coming from this spot spreads out the further out of focus one is. Under ideal conditions this produces a sort of "hourglass" shape of this point source in the third (axial) dimension. This shape is called the point spread function (PSF) of the microscope imaging system.

Since any fluorescence image is made up of a large number of such small fluorescent light sources the image is said to be "convolved by the point spread function".

Knowing this point spread function means, that it is possible to reverse this process to a certain extent by computer based methods commonly known as deconvolution microscopy. There are various algorithms available for 2D or 3D Deconvolution. They can be roughly classified in non restorative and restorative methods. While the non restorative methods can improve contrast by removing out of focus light from focal planes, only the restorative methods can actually reassign light to it proper place of origin. This can be an advantage over other types of 3D microscopy such as confocal microscopy, because light is not thrown away but reused. For 3D deconvolution one typically provides a series of images derived from different focal planes (called a Z-stack) plus the knowledge of the PSF which can be either derived experimentally or theoretically from knowing all contributing parameters of the microscope.

Sub-diffraction Optical Microscopy Techniques

It is well known that there is a spatial limit to which light can focus: approximately half of the wavelength of the light we are using. But this is not a true barrier, because this diffraction limit is only true in the far-field and localization precision can be increased with many photons and careful analysis (although two objects still cannot be resolved); and like the sound barrier, the diffraction barrier is breakable.

This part explores some approaches to imaging objects smaller than ~250 nm. Most of the following information was gathered (with permission) from a chemistry blog's review of sub-diffraction microscopy techniques.

TYPES OF MICROSCOPES

Various types of microscopes are available for use in the microbiology laboratory. The microscopes have varied applications and modifications that contribute to their usefulness.

The light microscope. The common light microscope used in the laboratory is called a compound microscope because it contains two types of lenses that function to magnify an object.

The lens closest to the eye is called the ocular, while the lens closest to the object is called the objective. Most microscopes have on their base an apparatus called a condenser, which condenses light rays to a strong beam. A diaphragm located on the condenser controls the amount of light coming through it. Both coarse and fine adjustments are found on the light microscope.

To magnify an object, light is projected through an opening in the stage, where it hits the object and then enters the objective. An image is created and this image becomes an object for the ocular lens, which remagnifies the image. Thus, the total magnification possible with the microscope is the magnification achieved by the objective multiplied by the magnification achieved by the ocular lens.

A compound light microscope often contains four objective lenses: the scanning lens (4X), the low-power lens (10X), the high-power lens (40 X) and the oil-immersion lens (100 X).

With an ocular lens that magnifies 10 times, the total magnifications possible will be 40 X with the scanning lens, 100 X with the low-power lens, 400 X with the high-power lens and 1000 X with the oil-immersion lens. Most microscopes are parfocal. This term means that the microscope remains in focus when one switches from one objective to the next objective.

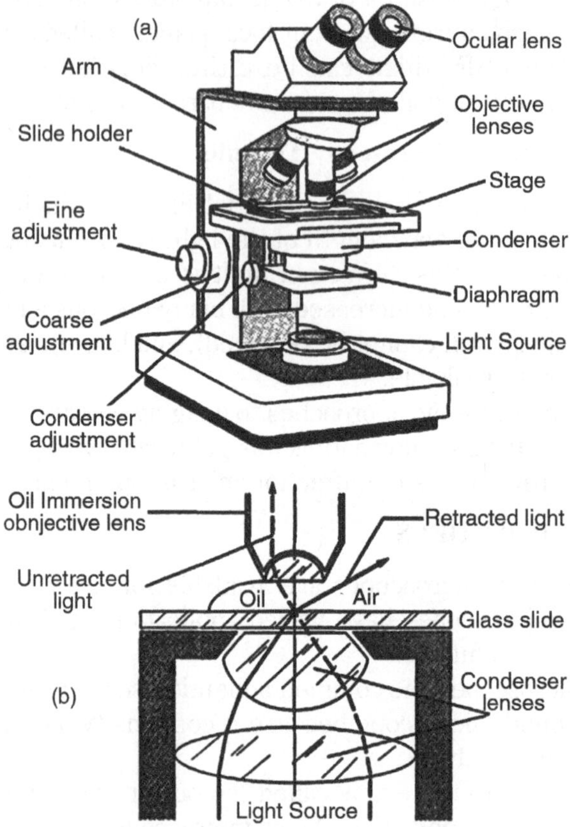

Fig. Light Microscopy.

(a) The Important Parts of a Common Light Microscope.

(b) How Immersion oil Gathers more Light for use in the Microscope

The ability to see clearly two items as separate objects under the microscope is called the resolution of the microscope. The resolution is determined in part by the wavelength of the light used for observing. Visible light has a wavelength of about 550 nm, while ultraviolet light has a wavelength of about 400 nm or less.

The resolution of a microscope increases as the wavelength decreases, so ultraviolet light allows one to detect objects not seen with visible light. The resolving power of a lens refers to the size of the smallest object that can be seen with that lens. The resolving power is based on the wavelength of the light used and the numerical aperture of the lens. The numerical aperture (NA) refers to the widest cone of light that can enter the lens; the NA is engraved on the side of the objective lens.

If the user is to see objects clearly, sufficient light must enter the objective lens. With modern microscopes, entry to the objective is not a problem for scanning, low-power and high-power lenses. However, the oil-immersion lens is exceedingly narrow and most light misses it. Therefore, the object is seen poorly and without resolution. To increase the resolution with the oil-immersion lens, a drop of immersion oil is placed between the lens and the glass slide.

Immersion oil has the same light-bending ability (index of refraction) as the glass slide, so it keeps light in a straight line as it passes through the glass slide to the oil and on to the glass of the objective, the oil-immersion lens. With the increased amount of light entering the objective, the resolution of the object increases and one can observe objects as small as bacteria. Resolution is important in other types of microscopy as well.

Other Light Microscopes

In addition to the familiar compound microscope, microbiologists use other types of microscopes for specific purposes. These microscopes permit viewing of objects not otherwise seen with the light microscope. An alternative microscope is the dark-field microscope, which is used to observe live spirochetes, such as those that cause syphilis. This microscope contains a special condenser that scatters light and causes it to reflect off the specimen at an angle. A light object is seen on a dark background.

A second alternative microscope is the phase-contrast microscope. This microscope also contains special condensers that throw light "out of phase" and cause it to pass through the object at different speeds. Live, unstained organisms are seen clearly with this microscope and internal cell parts such as mitochondria, lysosomes and the Golgi body can be seen with this instrument.

The fluorescent microscope uses ultraviolet light as its light source. When ultraviolet light hits an object, it excites the electrons of the object and they give off light in various shades of colour. Since ultraviolet light is used, the resolution of the object increases. A laboratory technique called the fluorescent-antibody technique employs fluorescent dyes and antibodies to help identify unknown bacteria.

Electron Microscopy

The energy source used in the electron microscope is a beam of electrons. Since the beam has an exceptionally short wavelength, it strikes most objects in its path and increases the resolution of the microscope significantly. Viruses and some large molecules can be seen with this instrument. The electrons travel in a vacuum to avoid contact with deflecting air molecules and magnets focus the beam on the object to be viewed. An image is created on a monitor and viewed by the technologist. The more traditional form of electron microscope is the transmission electron microscope (TEM).

To use this instrument, one places ultrathin slices of microorganisms or viruses on a wire grid and then stains them with gold or palladium before viewing. The densely coated parts of the specimen deflect the electron beam and both dark and light areas show up on the image. The scanning electron microscope (SEM) is the more contemporary form electron microscope. Although this microscope gives lower magnifications than the TEM, the SEM permits three-dimensional views of microorganisms and other objects. Whole objects are used and gold or palladium staining is employed.

Staining Techniques

Because microbial cytoplasm is usually transparent, it is necessary to stain microorganisms before they can be viewed with the light microscope. In some cases, staining is unnecessary, for example when microorganisms are very large or when motility is to be studied and a drop of the microorganisms can be placed directly on the slide and observed. A preparation such as this is called a wet mount.

A wet mount can also be prepared by placing a drop of culture on a cover-slip (a glass cover for a slide) and then inverting it over a hollowed-out slide. This procedure is called the hanging drop.

In preparation for staining, a small sample of microor ganisms is placed on a slide and permitted to air dry. The smear is heat fixed by quickly passing it over a flame. Heat fixing kills the organisms, makes them adhere to the slide and permits them to accept the stain.

Simple Stain Techniques

Staining can be performed with basic dyes such as crystal violet or methylene blue, positively charged dyes that are attracted to the negatively charged materials of the microbial cytoplasm. Such a procedure is the simple stain procedure. An alternative is to use a dye such as nigrosin or Congo red, acidic, negatively charged dyes. They are repelled by the negatively charged cytoplasm and gather around the cells, leaving the cells clear and unstained. This technique is called the negative stain technique.

Differential Stain Techniques

The differential stain technique distinguishes two kinds of organisms. An example is the Gram stain technique. This differential technique separates bacteria into two groups, Gram-positive bacteria and Gram-negative bacteria. Crystal violet is first applied, followed by the mordant iodine, which fixes the stain.

Then the slide is washed with alcohol and the Gram-positive bacteria retain the crystal-violet iodine stain; however, the Gram-negative bacteria lose the stain. The Gram-negative bacteria subsequently stain with the safranin dye, the counterstain, used next. These bacteria appear red under the oil-immersion lens, while Gram-positive bacteria appear blue or purple, reflecting the crystal violet retained during the washing step.

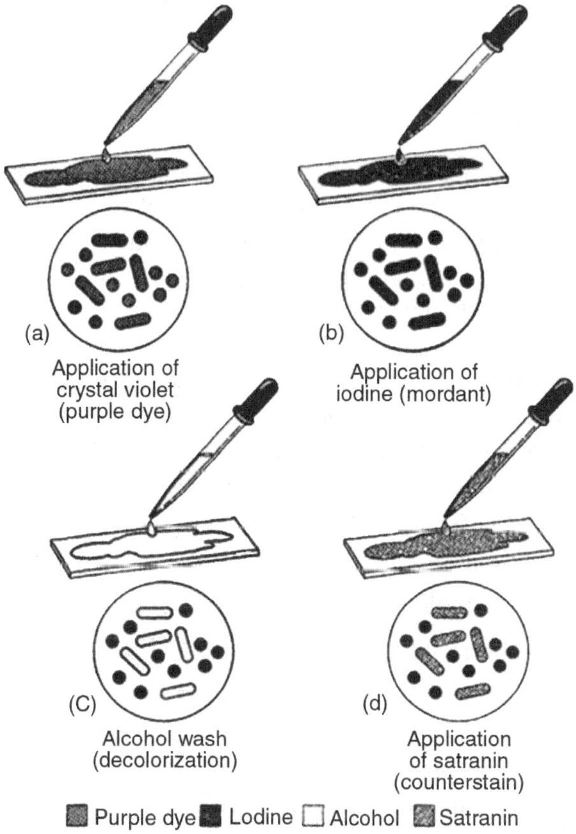

Fig. The Gram Stain Procedure used for Differentiating Bacteria into two Groups

Another differential stain technique is the acid-fast technique. This technique differentiates species of *Mycobac-terium* from other bacteria. Heat or a lipid solvent is used to carry the first stain, carbolfuchsin, into the cells. Then the cells are washed with a dilute acid-alcohol solution. *Mycobacterium* species resist the effect of the acid-alcohol and retain the carbolfuchsin stain (bright red). Other bacteria lose the stain and take on the subsequent methylene blue stain (blue). Thus, the acid-fast bacteria appear bright red, while the nonacid-fast bacteria appear blue when observed under oil-immersion microscopy.

Other stain techniques seek to identify various bacterial structures of importance. For instance, a special stain technique highlights the flagella of bacteria by coating the flagella with dyes or metals to increase their width. Flagella so stained can then be observed.

A special stain technique is used to examine bacterial spores. Malachite green is used with heat to force the stain into the cells and give them colour. A counterstain, safranin, is then used to give colour to the nonsporeforming bacteria. At the end of the procedure, spores stain green and other cells stain red.

LIGHT MICROSCOPY

The light microscope, so called because it employs visible light to detect small objects, is probably the most well-known and well-used research tool in biology. Yet, many students and teachers are unaware of the full range of features that are available in light microscopes. Since the cost of an instrument increases with its quality and versatility, the best instruments are, unfortunately, unavailable to most academic programs. However, even the most inexpensive "student" microscopes can provide spectacular views of nature and can enable students to perform some reasonably sophisticated experiments.

A beginner tends to think that the challenge of viewing small objects lies in getting enough magnification. In fact, when it comes to looking at living things the biggest challenges are, in order,

- Obtaining sufficient contrast
- Finding the focal plane
- Obtaining good resolution
- Recognizing the subject when one sees it

The smallest objects that are considered to be living are the bacteria. The smallest bacteria can be observed and cell shape recognized at a mere 100x magnification. They are invisible in bright field microscopes, though.

Types of Light Microscopes

The bright field microscope is best known to students and is most likely to be found in a classroom. Better equipped classrooms and labs may have dark field and/or phase contrast optics. Differential interference contrast, Nomarski, Hoffman modulation contrast and variations produce considerable depth of resolution and a three dimensional effect. Fluorescence and confocal microscopes are specialized instruments, used for research, clinical and industrial applications.

Other than the compound microscope, a simpler instrument for low magnification use may also be found in the laboratory. The stereo microscope, or dissecting microscope usually has a binocular eyepiece tube, a long working distance and a range of magnifications typically from 5x to 35 or 40x. Some instruments supply lenses for higher magnifications, but there is no improvement in resolution. Such "false magnification" is rarely worth the expense.

Bright Field Microscopy

With a conventional bright field microscope, light from an incandescent source is aimed toward a lens beneath the stage called the condenser, through the specimen, through an objective lens and to the eye through a second magnifying lens, the ocular or eyepiece. We see objects in the light path because natural pigmentation or stains absorb light differentially, or because they are thick enough to absorb a significant amount of light despite being colorless.

A *Paramecium* should show up fairly well in a bright field microscope, although it will not be easy to see cilia or most organelles. Living bacteria won't show up at all unless the viewer hits the focal plane by luck and distorts the image by using maximum contrast.

A good quality microscope has a built-in illuminator, adjustable condenser with aperture diaphragm (contrast) control, mechanical stage and binocular eyepiece tube.

The condenser is used to focus light on the specimen through an opening in the stage. After passing through the specimen, the light is displayed to the eye with an apparent field that is much larger than the area illuminated.

The magnification of the image is simply the objective lens magnification (usually stamped on the lens body) times the ocular magnification. Students are usually aware of the use of the coarse and fine focus knobs, used to sharpen the image of the specimen.

They are frequently unaware of adjustments to the condenser that can affect resolution and contrast. Some condensers are fixed in position, others are focusable, so that the quality of light can be adjusted. Usually the best position for a focusable condenser is as close to the stage as possible.

The bright field condenser usually contains an aperture diaphragm, a device that controls the diameter of the light beam coming up through the condenser, so that when the diaphragm is stopped down (nearly closed) the light comes straight up through the centre of the condenser lens and contrast is high.

When the diaphragm is wide open the image is brighter and contrast is low. A disadvantage of having to rely solely on an aperture diaphragm for contrast is that beyond an optimum point the more contrast we produce the more we distort the image. With a small, unstained, unpigmented specimen, we are usually past optimum contrast when we begin to see the image.

Mount the Specimen on the Stage

The cover slip must be up if there is one. High magnification objective lenses can't focus through a thick glass slide; they must be brought close to the specimen, which is why coverslips are so thin. The stage may be equipped with simple clips (less expensive microscopes), or with some type of slide holder. The slide may require manual positioning, or there may be a mechanical stage (preferred) that allows precise positioning without touching the slide.

Optimize the Lighting

A light source should have a wide dynamic range, to provide high intensity illumination at high magnifications and lower intensities so that the user can view comfortably at low magnifications.

Better microscopes have a built-in illuminator and the best microscopes have controls over light intensity and shape of the light beam. If our microscope requires an external light source, make sure that the light is aimed toward the middle of the condenser. Adjust illumination so that the field is bright without hurting the eyes.

Adjust the Condenser

To adjust and align the microscope, start by reading the manual. If no manual is available, try using these guidelines.

If the condenser is focusable, position it with the lens as close to the opening in the stage as we can get it.

If the condenser has selectable options, set it to bright field. Start with the aperture diaphragm stopped down (high contrast). We should see the light that comes up through the specimen change brightness as we move the aperture diaphragm lever.

FOCUS, LOCATE AND CENTRE THE SPECIMEN

Start with the lowest magnification objective lens, to home in on the specimen and/or the part of the specimen we wish to examine. It is rather easy to find and focus on sections of tissues, especially if they are fixed and stained, as with most prepared slides.

However it can be very difficult to locate living, minute specimens such as bacteria or unpigmented protists. A suspension of yeast cells makes a good practice specimen for finding difficult objects.

- Use dark field mode (if available) to find unstained specimens. If not, start with high contrast (aperture diaphragm closed down).
- Start with the specimen out of focus so that the stage and objective must be brought closer together. The first surface to come into focus as we bring stage and objective together is the top of the cover slip. With smears, a cover slip is frequently not used.
- If we are having trouble, focus on the edge of the cover slip or an air bubble, or something that we can readily recognize. The top edge of the cover slip comes into focus first, then the bottom, which should be in the same plane as our specimen.
- Once we have found the specimen, adjust contrast and intensity of illumination and move the slide around until we have a good area for viewing.

Adjust Eyepiece Separation, Focus

With a single ocular, there is nothing to do with the eyepiece except to keep it clean. With a binocular microscope (preferred) we need to adjust the eyepiece separation just like we do a pair of binoculars. Binocular vision is much more sensitive to light and detail than monocular vision, so if we have a binocular microscope, take advantage of it.

One or both of the eyepieces may be a telescoping eyepiece, that is, we can focus it. Since very few people have eyes that are perfectly matched, most of us need to focus one eyepiece to match the other image. Look with the appropriate eye into the fixed eyepiece and focus with the microscope focus knob. Next, look into the adjustable eyepiece (with the other eye of course) and adjust the eyepiece, not the microscope.

Select an Objective lens for Viewing

The lowest power lens is usually 3.5 or 4x and is used primarily for initially finding specimens. We sometimes call it the scanning lens for that reason. The most frequently used objective lens is the 10x lens, which gives a final magnification of 100x with a 10x ocular lens.

For very small protists and for details in prepared slides such as cell organelles or mitotic. Typical high magnification lenses are 40x and 97x or 100x. The latter two magnifications are used exclusively with oil in order to improve resolution.

Move up in magnification by steps. Each time we go to a higher power objective, re-focus and re-centre the specimen. Higher magnification lenses must be physically closer to the specimen itself, which poses the risk of jamming the objective into the specimen.

Be very cautious when focusing. By the way, good quality sets of lenses are parfocal, that is, when we switch magnifications the specimen remains in focus or close to focused. Bigger is not always better. All specimens have three dimensions and unless a specimen is extremely thin as suggested, be unable to focus with a high magnification objective. The higher the magnification, the harder it is to "chase" a moving specimen. The apparent field of an eyepiece is constant regardless of magnification used.

So it follows that when we raise magnification the area of illuminated specimen we see is smaller. Since we are looking at a smaller area, less light reaches the eye and the image darkens. With a low power objective we may have to cut down on illumination intensity. With a high power we need all the light we can get, especially with less expensive microscopes.

Bright field microscopy is best suited to viewing stained or naturally pigmented specimens such as stained prepared slides of tissue sections or living photosynthetic organisms. It is useless for living specimens of bacteria and inferior for non-photosynthetic protists or metazoans, or unstained cell suspensions or tissue sections. Here is a not-so-complete list of specimens that might be observed using bright-field microscopy and appropriate magnifications (preferred final magnifications are emphasized).

- *Prepared slides, stained:* Bacteria (1000x), thick tissue sections (100x, 400x), thin sections with condensed chromosomes or specially stained organelles (1000x), large protists or metazoans (100x).
- *Smears, stained:* Blood (400x, 1000x), negative stained bacteria (400x, 1000x).
- *Living preparations (wet mounts, unstained):* Pond water (40x, 100x, 400x), living protists or metazoans (40x, 100x, 400x occasionally), algae and other microscopic plant material (40x, 100x, 400x). Smaller specimens will be difficult to observe without distortion, especially if they have no pigmentation.

Care of the Microscope

- Everything on a good quality microscope is unbelievably expensive, so be careful.
- Hold a microscope firmly by the stand, only. Never grab it by the eyepiece holder, for example.
- Hold the plug (not the cable) when unplugging the illuminator.
- Since bulbs are expensive and have a limited life, turn the illuminator off when we are done.
- Always make sure the stage and lenses are clean before putting away the microscope.
- Never use a paper towel, a kimwipe, our shirt, or any material other than good quality lens tissue or a cotton swab (must be 100% natural cotton) to clean an optical surface. We may use an appropriate lens cleaner or distilled water to help remove dried material.
- Organic solvents may separate or damage the lens elements or coatings.
- Cover the instrument with a dust jacket when not in use.
- Focus smoothly; don't try to speed through the focusing process or force anything. For example if we encounter increased resistance when focusing then we 've probably reached a limit and we are going in the wrong direction.

INDUSTRIAL MICROBES

The enslavement and death of countless millions of workers may result from a newly discovered method of producing glycerine (currently produced as a byproduct of soap manufacture). Whether the process reported by Canada's National Research Council in Ottawa will be adopted commercially depends largely on the efficiency of the workers' digestive apparatus. In any case, no one will protest their exploitation. The workers are microscopic members of the clan Bacillus subtilis.

The use of microbes as minuscule chemical factories has been practiced, if not understood, since the first butter was churned, the first wine pressed, the first beer brewed. Spurred by advances in the field of biochemistry and the pressures of two wars, the employment of microbe labour has recently spread to a whole new field of chemical manufacture. Little is known of the actual metabolic process by which microbes work, but by careful control and endless experiment chemists and biologists, working together, have been able to set them hundreds of chemical tasks.

Best known of these industrial microorganisms is yeast, whose appetite for the carbohydrates in beets, sugar cane, wood and other fibrous vegetable matter made possible the production in 1944 of about 638 million gallons of alcohol—grain and wood. But the yeasts are only one group of the microbic multitude able to perform specific jobs. Bacteria resembling the bacilli of human ailments and molds like mildew have also been put to work in industry. Some microbes have done double jobs. One, fed in a certain way, yields oxalic acid, basic chemical of the blueprint industry; on a different diet it produces the gluconic acid used in medicines. The versatile Clostridium acetobutylicum, on a single diet of corn mash, produces acetone for solvents, butanol for automobile lacquers and riboflavin (Vitamin 62).

Grasshoppers and Cantaloupes. In a constant search for new microscopic workers, industrial, university and Government researchers have isolated over 200,000 varieties. Useful microbes may turn up anywhere—in the air, on the water, on forest leaf mold, in city garbage cans. A potent industrial bacillus was discovered in the intestines of a grasshopper. The best strain of the mold Penicillium notatum, which makes life-saving penicillin, was first noticed on a cantaloupe rind. Despite their humble origins, microbes are as temperamental as coloraturas.

They are fussy about temperature, about food and about the company they keep. Some like plenty of air, some like none and some govern their behaviour according to whether they get it or not. In an airless, quiet place, yeast will produce wine; in air it just reproduces itself. To keep such un reliable workers healthy, happy and productive is the responsibility of a growing new species of industrial scientist, the biological engineer. His job: to reproduce on a factory scale biological processes and conditions none too easy to control in a laboratory test tube.

Biological Engineering. Leaders in the development of this field have been technicians of such firms as Brooklyn's Charles Pfizer & Co., one of the oldest of the fermentation chemical makers and a big producer of penicillin. They have devised techniques that smack less of a factory than of the contagion ward of a hospital.

The production of penicillin, which is typical of most fermentation processes, starts from a small culture of the mold growing on a mixture of water, sugar and some nitrogenous food. A 10,000-gal. vat of nutrient may be charged with as much as 100 gal. of this mixture, which has been kept sterilized so that no impurity can possibly be present. Should another, different microbe find his way in, the batch would be spoiled.

In theory, all that is necessary to start production is to introduce the microscopic microbe into the large vat. In practice, the vats are inoculated with the larger cultures, to save time, but at the end of three days the biological engineer may find the clear, brownish liquid filled with the clusters of strawlike gold that he had hoped to produce. Or he may find only an accumulation of horrible smells.

Microbe Mysteries. The biologist is often completely in the dark as to why his microbes fail so dismally, when they do. Cross contamination of one culture by another is a common cause.

To prevent it at Pfizer's, no two chemicals are made in the same building. Workers with coughs or sneezes are kept away and those going in & out are thoroughly sterilized and dressed like hospital interns. Filtered air and ultra violet light are constantly at work.

In spite of all precautions, the microbes find ways to flout their keepers. A simon-pure culture may sometimes, apparently from sheer perversity, change its nature. After such a mutation, it may produce a better product, a worse product, or an entirely different product. Biologists are no more certain of the reason for this than they are of the nature of the sub-microscopic organisms which plague the microbes themselves. Absenteeism among microbe workers is no problem, but in at least one factory devoted to the manufacture of essential chemical solvents, all work had to be stopped for days because the microbes themselves got sick.

PRODUCTION AND USE OF MICROBIAL ENZYMES FOR DAIRY PROCESSING

Indians are known to be lovers of milk and its products. As a sequel to white revolution, India has surged ahead to become the largest milk producer in the world, the production figure touched to 74 million tons for the year 1998. With growing urbanization, demand for processed dairy foods has increased considerably, in particular demand for different cheese varieties and low-lactose milk due to increasing intolerance of human beings to lactose in milk and other milk products. For improving the quality of milk and milk products, a number of different enzymes from microbial as well as from nonmicrobial sources have potential applications in dairy processing. The use of rennet in cheese manufacture was among the earliest applications of exogenous enzymes in food processing, dating back to approximately 6000 B C. In 1994, the total production of cheese was 8000 metric tons against a total demand of 9000 metric tons.

The projected demand by 2000 A D is around 30,000 metric tons. The use of rennet, as an exogenous enzyme, in cheese manufacture is perhaps the largest single application of enzymes in food processing. In recent years, proteinases have found additional applications in dairy technology, for example in acceleration of cheese ripening, modification of functional properties and preparation of dietic products. Principal among some enzymes that have important and growing applications are lipases and b-galactosidases.

Enzymes with limited applications include glucose oxidase, superoxide dismutase, sulphydryl oxidase, etc. The increasing use of enzymes to produce specific products with characteristic attributes can be emphasized by the world-wide sale of industrial enzymes approximating to US $ 1.6 billion, which is expected to reach US $ 3.0 billion by the year 2008. Almost 45% of this is shared by the food industry and the remaining is shared by detergent (34.4%), textile (11%), leather (2.8%), pulp and paper (1.2%) industries and other industries (5.6%) excluding enzymes for use in diagnostics and therapeutics.

India being the highest producer of milk in the world and consequently the surplus availability of milk in our country has triggered the food and dairy industry to convert the liquid milk into value-added products using biochemical and enzymatic processes.

Microbial Rennets in Dairy Applications

Animal rennet (bovine chymosin) is conventionally used as a milk-clotting agent in dairy industry for the manufacture of quality cheeses with good flavour and texture. Owing to an increase in demand for cheese production world wide – i.e. 4% per annum over the past 20 years, approximating 13.533 million tons – coupled with reduced supply of calf rennet, has therefore led to a search for rennet substitutes, such as microbial rennets. At present, microbial rennet is used for one-third of all the cheese produced world wide.

Rennin acts on the milk protein in two stages, by enzymatic and by nonenzymatic action, resulting in coagulation of milk. In the enzymatic phase, the resultant milk becomes a gel due to the influence of calcium ions and the temperature used in the process. Many microorganisms are known to produce rennet-like proteinases which can substitute the calf rennet. Microorganisms like Rhizomucor pusillus, R. miehei, Endothia parasitica, Aspergillus oryzae and Irpex lactis are used extensively for rennet production in cheese manufacture. Extensive research that has been carried out so far on rennet substitutes has been reviewed by several authors. Different strains of species of Mucor are often used for the production of microbial rennets.

Whereas best yields of the milk-clotting protease from Rhizomucor pusillus are obtained from semisolid cultures containing 50% wheat bran, R. miehei and Endothia parasitica are well suited for submerged cultivation. Using the former, good yields of milk-clotting protease may be obtained in a medium containing 4% potato starch, 3% soybean meal and 10% barley. During growth, lipase is secreted together with the protease. Therefore, the lipase activity has to be destroyed by reducing the pH, before the preparation can be used as cheese rennet.

Microbial rennets from various microorganisms (marketed under the trade names such as Rennilase, Fromase, Marzyme, Hanilase, etc.) being marketed since the 1970s have proved satisfactory for the production of different kinds of cheese. The molecular and enzymatic properties of chymosins have been studied extensively.

Although the proteolytic specificities of the three commonly used fungal rennets are considerably different from those of calf chymosin, these rennets have been used to produce acceptable cheeses. Recently Novo Nordisk has succeeded in expressing just one proteolytic enzyme from the fungus R. miehei in the well-known organism A. oryzae.

This host organism is able to produce the single protease that cleaves the casein into a glycomacropeptide and para casein by hydrolysing only at the phe105–met106 peptide bond between phenyl alanine and methionine. This monocomponent enzyme product has the trade name Novoren. One major drawback of microbial rennet use in cheese manufacture, is the development of off flavour and bitter taste in the nonripened as well as in the ripened cheeses. The rennets from microbial sources are more proteolytic in nature in comparison to rennet from animal sources, resulting in production of some bitter peptides during the process of cheese ripening. Hence, attempts have been made to clone the gene for calf chymosin and to express it in selected bacteria, yeasts and molds.

Recombinant Rennets for Cheese Manufacture

Due to shortage of calf stomachs and the economic value of cheese rennet, gene for calf chymosin was one of the first genes for mammalian enzymes that was cloned and expressed in microorganisms. Many different laboratories have cloned the gene for calf prochymosin in Escherichia coli and analysed the structure of the gene as well as the properties of the recombinant chymosin. The expressed proenzyme in E. coli, is present mainly as insoluble inclusion bodies, comprised of reduced prochymosin as well as molecules that are interlinked by disulphide bridges. After disintegration of the cells, inclusion bodies are harvested by centrifugation.

The individual laboratories have reported some differences in the procedure for renaturation of prochymosin from the inclusion bodies, but all have followed the same general scheme. The enzymatic properties of recombinant E. coli chymosin are indistinguishable from those of native calf chymosin. The enzymes were identical when observed by immunodiffusion in gels, but a slight difference was observed by enzyme linked immunosorbent assay (ELISA).

Table. Recombinal Chymosin Proparations at/or Approaching Legal and Commercial Acceptance.

Sourcc of DNA microorganisms	Praducing (brand namc)	Prducing company
Calf abomasums (Maxiren)	klayveromyces lactis	Gist Brocades
Calf abomasums Hansen	Aspergillus niger	Genccor/chr. (Chyrnoger)
Synthetic	Escherichic coli	Pfizer (Chymax)

The gene for prochymosin has also been cloned in Saccharomyces cerevisiae; the levels of expression have been reported to be 0.5 to 2.0% of total yeast protein. In yeast, about 20% of the prochymosin can be released in soluble form which can be activated directly; the remaining 80% is still associated with the cell debris.

The cloning of chymosin was carried out without the prosequence. Though the level of chymosin mRNA was similar to that of prochymosin mRNA, no milk-clotting activity was observed in clones containing the chymosin gene only.

The results suggest that the prosequence is essential for correct folding of the polypeptide chain. The zymogen for the aspartic proteases from R. pusillus, also called mucor rennin, has likewise been cloned and expressed in yeast. Studies on its conversion to active form showed that secretion of R. pusillus protease from recombinant yeast was dependent on glycosylation of the enzyme.

Compared to yeast, the filamentous fungi generally secrete larger quantities of proteins into the culture medium. Furthermore, filamentous fungi secrete the heterologous proteins with correct folding of the polypeptide chain and process correct pairing of sulphydryl groups.

The gene for R. michci protcase has been expressed in A. oryzae. Prochymosin has been expressed in Kluyveromyces lactis, A. nidulans, A. niger and Trichoderma rccsci. In most of the cases, the reported yields of the model systems were about 10–40 mg of enzyme per litre of culture medium. However, 3.3 g of enzyme per liter has also been achieved. Yeast, Kluyveromyces lactis, has recently been used as an efficient host for the secretion of recombinant chymosin, which has led to the development of large-scale production process for chymosin. If produced on an industrial scale, the yields will perhaps be of the latter magnitude. Chymosins which are approaching or have marketing and legal acceptance and generally regarded as safe (GRAS). Most of the companies produce recombinant rennet of cattle calf origin in different microbial hosts, however in India, the major source of milk is buffalo, showing a different composition from that of cow. The rennet from buffalo source has inherent compatibility for clotting buffalo milk.

Therefore, National Dairy Research Institute, Karnal, has taken a lead in cloning gene of buffalo chymosin. The buffalo chymosin cDNA has been cloned in E. coli and partial N-terminal sequencing of the purified buffalo chymosin indicates that it is highly homologous to the cattle chymosin.

Several cheese making experiments have been carried out with recombinant chymosin and the general aspects of recombinant chymosin have been dealt with in a report. Since most of the rennet (> 90 %) added to cheese milk is lost in the whey, immobilization would considerably extend its catalytic life. Several rennets have been immobilized, but their efficiency as milk coagulants has been questioned.

So, there is a fairly general support for the view that immobilized enzymes cannot coagulate milk properly owing to inaccessability of the Phe–Met peptide bond of k-casein and that the apparent coagulating activity of immobilized rennets is due to leaching of the enzyme from the support. Different types of conventional cheeses have been successfully made by using recombinant rennet on an experimental or pilot scale.

No major differences have been detected between cheeses made with recombinant chymosins or natural enzymes, regarding cheese yield, texture, smell, taste and ripening. The recombinant chymosins are identical with calf rennet according to the report on biochemical and genetic evidences.

Lactase in Dairy Industry

Lactose, the sugar found in milk and whey and its corresponding hydrolase, lactase or b -galactosidase, have been extensively researched during the past decade. This is because of the enzyme immobilization technique which has given new and interesting possibilities for the utilization of this sugar.

Because of intestinal enzyme insufficiency, some individuals and even a population, show lactose intolerance and difficulty in consuming milk and dairy products.

Hence, low-lactose or lactose-free food aid programme is essential for lactose-intolerant people to prevent severe tissue dehydration, diarrhoea and, at times, even death. Another advantage of lactase-treated milk is the increased sweetness of the resultant milk, thereby avoiding the requirement for addition of sugars in the manufacture of flavoured milk drinks. Manufacturers of ice cream, yoghurt and frozen desserts use lactase to improve scoop and creaminess, sweetness and digestibility and to reduce sandiness due to crystallization of lactose in concentrated preparations. Cheese manufactured from hydrolysed milk ripens more quickly than the cheese manufactured from normal milk.

Technologically, lactose crystallizes easily which sets limits to certain processes in the dairy industry and the use of lactase to overcome this problem has not reached its fullest potential because of the associated high costs. Moreover, the main problem associated with discharging large quantities of cheese whey is that it pollutes the environment. But, the discharged whey could be exploited as an alternate cheap source of lactose for the production of lactic acid by fermentation.

The whey permeate, which is a by-product in the manufacture of whey protein concentrates, by ultrafiltration could be fermented efficiently by Lactobacillus bulgaricus. Lactose can be obtained from various sources like plants, animal organs, bacteria, yeasts (intracellular enzyme), or molds. Some of these sources are used for commercial enzyme preparations.

Lactase preparations from A. niger, A. oryzae and Kluyveromyces lactis are considered safe because these sources already have a history of safe use and have been subjected to numerous safety tests. The most investigated E. coli lactase is not used in food processing because of its cost and toxicity problems.

Properties of Lactase

The properties of the enzyme depend on its source. Temperature and pH optima differ from source to source and with the type of particular commercial preparation.

Immobilization of the enzymes, method of immobilization and type of carrier can also influence these optima values. In general, fungal lactase have pH optima in the acidic range 2.5–4.5 and yeast and bacterial lactases in the neutral region 6–7 and 6.5–7.5, respectively.

The variation in pH optima of lactases makes them suitable for specific applications, for example fungal lactases are used for acid whey hydrolysis, while yeast and bacterial lactases are suitable for milk (pH 6.6) and sweet whey (pH 6.1) hydrolysis. Product inhibition, e.g. inhibition by galactose, is another property which also depends on the source of lactase.

The enzyme from A. niger is more strongly inhibited by galactose than that from A. oryzae. This inhibition can be overcome by hydrolysing lactose at low concentrations by using immobilized enzyme systems or by recovering the enzyme using ultrafiltration after batch hydrolysis.

Lactase from Bacillus species are superior with respect to thermostability, pH operation range, product inhibition and sensitivity against high-substrate concentration.

Thermostable enzymes, able to retain their activity at 60°C or above for prolonged periods, have two distinct advantages viz. they give higher conversion rate or shorter residence time for a given conversion rate and the process is less prone to microbial contamination due to higher operating temperature. Bacillus species have a pH optima of 6.8 and a temperature optima of 65°C. Its high activity for skim milk and less inhibition by galactose has made it suitable for use as a production organism for lactase.

The enzymatic hydrolysis of lactose can be achieved either by free enzymes, usually in batch fermentation process, or by immobilized enzymes or even by immobilized whole cells producing intracellular enzyme. Although numerous hydrolysis systems have been investigated, only few of them have been scaled up with success and even fewer have been applied at an industrial or semi-industrial level.

Several acid hydrolysis systems have been developed to industry-scale level. Large-scale systems which use free enzyme process have been developed for processing of UHT-milk and processing of whey, using K. lactis lactase (Maxilact, Lactozyme).

Several commercial immobilized systems have been developed for commercial exploitation. Snamprogretti process of industrial-scale milk processing technology in Italy is one such working systems.

They make use of fibre-entrapped yeast lactase in a batch process and the milk used is previously sterilized by UHT. For pilot plants, there are three other processes designed and developed to handle milk:

- Gist-Brocades, Rohm GmbH (Germany);
- Sumitomo, Japan. These are continuous processes with short residence times. Processing of whey UF-permeate is accomplished by the system developed by Corning Glass, Connecticut, Lehigh, Valio and Amerace corp. The process by Corning Glass is being applied at commercial scale in the bakers yeast production using hydrolysed whey.

Microbial Enzymes in Accelerated Cheese Ripening

Cheese ripening is a complex process mediated by biochemical and biophysical changes during which a bland curd is developed into a mature cheese with characteristic flavour, texture and aroma. The desirable attributes are produced by the partial and gradual breakdown of carbohydrates, lipids and proteins during ripening, mediated by several agents, viz.

- Residual coagulants,
- Starter bacteria and their enzymes,
- Nonstarter bacteria and their enzymes,
- Indegenous milk enzymes, especially proteinases;
- Secondary inocula with their enzymes.

Proteolysis occurs in all the cheese varieties and is a prerequisite for characteristic flavour development that can be regulated by proper use of the above agents. Cheese ripening is essentially an enzymatic process which can be accelerated by augmenting activity of the key enzymes. This has the advantage of initiating more specific action for flavour development compared to use of elevated temperatures that can result in accelerating undesirable nonspecific reactions and consequently off flavour development.

Enzymes may be added to develop specific flavours in cheeses, for example lipase addition for the development of Parmesan or Blue-type cheese flavours. Attempts to accelerate the multiple secondary flavour-forming-reactions, e.g. Strecker degradation, have been scarce.

The pathways leading to the formation of flavour compounds are largely unknown and therefore the use of exogenous enzymes to accelerate ripening is mostly an empirical process. Different microbial enzymes used to accelerate cheese ripening.

Studies on the hydrolysis of whole casein and isolated casein components to observe the kinetics and specificity of aspartic proteases in rennin, pepsin and four microbial rennet substitutes indicated the considerable differences in the reaction velocity and the extent of hydrolysis on the rennet curd yield. The rennet enzymes were active even in the later phases of cheese-ripening.

Proteinases and Peptidases

Proteolysis is characteristic of most cheese varieties and is indispensable for good flavour and textural development. Proteinases used in cheese processing include:

- Plasmin,
- Rennet;
- Proteinases (cell wall and/or intracellular) of the starter and nonstarter bacteria.

Approximately 6% of the rennet added to cheese milk remains in the curd after manufacture and contributes significantly to proteolysis during ripening. Combinations of individual neutral proteinases and microbial peptidases intensified cheese flavour and when used in combinations with microbial rennets reduced the intensity of bitterness caused by the latter. Acid proteases in isolation cause intense bitterness.

Various animal or microbial lipases gave pronounced cheese flavour, low bitterness and strong rancidity, while lipases in combination with proteinases and/or peptidases give good cheese flavour with low levels of bitterness. In a more balanced approach to the acceleration of cheese ripening using mixtures of proteinases and peptidases, attenuated starter cells or cell-free extracts (CFE) are being favoured.

Table. Microbial Enzymes used to Ucoclcrate Chocse Ripening

Enzyme	Source
Microbial lipases	Aspergillus niger A. oryzae Rhizomucor miehei
Lactasce	Streptococcus lactis
Ktuyveromyces sp.	
	Escherichia coli
Microbial serline proteinases	A. niger
Neutral pmroteinases	Bacillus subtilis A. oryzae

PROTEOLYTIC ENZYMES OF LACTIC ACID BACTERIA IN FERMENTED MILK PRODUCTS

The proteolytic system of lactic acid bacteria is essential for their growth in milk and contributes significantly to flavour development in fermented milk products.

The proteolytic system is composed of proteinases which initially cleaves the milk protein to peptides; peptidases which cleave the peptides to small peptides and amino acids; and transport system responsible for cellular uptake of small peptides and amino acids.

Lactic acid bacteria have a complex proteolytic system capable of converting milk casein to the free amino acids and peptides necessary for their growth. These proteinases include extracellular proteinases, endopeptidases, aminopeptidases, tripeptidases and proline-specific peptidases, which are all serine proteases. Apart from lactic streptococcal proteinases, several other proteinases from nonlactostreptococcal origin have been reported.

There are also serine type of proteinases, e.g. proteinases from Lactobacillus acidophillus, L. plantarum, L. delbrueckii sp. bulgaricus, L. lactis and L. helveticus. Aminopeptidases are important for the development of flavour in fermented milk products, since they are capable of releasing single amino acid residues from oligopeptides formed by extracellular proteinase activity.

Other Dairy Enzymes

Other enzymes used for dairy food application include:

- Proteases to reduce allergic properties of cow milk products for infants;
- Lipases for development of lipolytic flavours in speciality cheeses.

The functional properties of milk proteins may be improved by limited proteolysis through the enzymatic modification of milk proteins.

An acid-soluble casein, free of off flavour and suitable for incorporation into beverages and other acid foods, has been prepared by limited proteolysis.

The antigenicity of casein is destroyed by proteolysis and the hydrolysate is suitable for use in milk-protein-based foods for infants allergic to cow milk. Lipolysis makes an important contribution to swiss cheese flavours, due mainly to the lipolytic enzymes of the starter cultures.

The characteristic peppery flavour of Blue cheese is due to short-chain fatty acids and methyl ketones. Most of the lipolysis in Blue cheese is catalysed by Penicillium roqueforti lipase, with a lesser contribution from indigenous milk lipase.

The NOVO process for production of enzyme modified cheese (EMC) uses medium-aged cheese which is emulsified, homogenized and pasteurized, after which 'palatase' (a lipase from R. miehei) is added, with or without a proteinase and the blend is ripened at a high temperature for one to four days.

The mixture is reheated, a paste results which is suitable for inclusion in soups, dips, dressings, or snack foods. EMC technology has been developed to produce a range of characteristic cheese flavours and flavour intensities, for example swiss, blue, cheddar, provolo-nemor or romano, suitable for inclusion at low levels in many products.

The claims that exogenous enzymes are effective in accelerating ripening have not led to their wide-spread use, possibly due to their high cost, difficulties in distributing them uniformly in the curd and the possible danger of over-ripening the cheese. The other minor enzymes having limited applications in dairy processing include glucose oxidase, catalase, superoxide dismutase, sulphydryl oxidase, lactoperoxidase and lysozymes.

Glucose oxidase and catalase are often used together in selected foods for preservation. Superoxide dismutase is an antioxidant for foods and generates H_2O_2, but is more effective when catalase is present. Thermally induced generation of volatile sulphydryl groups is thought to be responsible for the cooked off-flavour in ultra high temperature (UHT) processed milk. Use of sulphydryl oxidase under aseptic conditions can eliminate this defect. The natural inhibitory mechanism in raw milk is due to the presence of low levels of lactoperoxidase (LP), which can be activated by the external addition of traces of H_2O_2 and thiocyanate.

It has been reported that the potential of LP-system and its activation enhances the keeping quality of milk. Cow milk can be provided with protective factors by the addition of lysozyme, making it suitable as an infant milk. Lysozyme acts as a preservative by reducing bacterial counts in milk without affecting the The scope of application of minor enzymes to milk and milk products has been recently reviewed.

The global market for the production of microbial enzymes for use in dairy-products manufacture is considerably large, but is being dominated only by a limited number of enzyme producers. In India the microbial dairy enzymes requirement has been very limited till now.

However, with the advent of technological processes for the manufacture of different varieties of milk products, such as cheeses by the State Dairy Federations, Co-operatives and Private Dairy Product Manufacturers like Amul, Vijaya, Verka, Dynamix, Nestle, Smith Kline Beechem, etc., the markets for the sale of such products in megacities and towns has been slowly growing for the past two to three years.

Presently, many of these microbial enzymes, such as microbial rennets and other enzymes, are being imported. Hence, there is a scope for the production of enzymes such as microbial rennet, lactase, proteinases and lipases indigenously. In the near future, the requirement for these enzymes is bound to increase by leaps and bounds, basically due to requirement of value-added dairy products in the country.

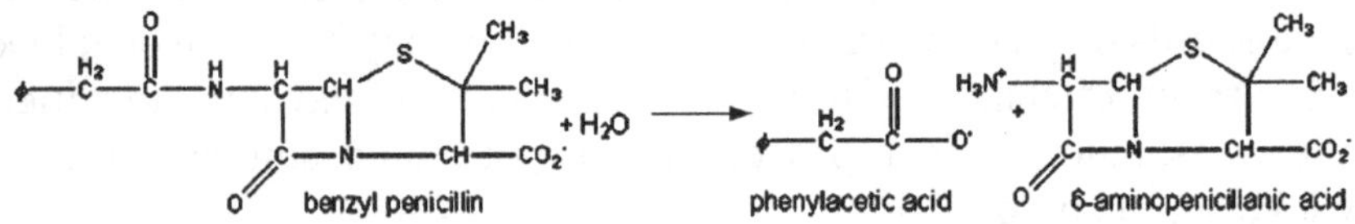

PRODUCTION OF ANTIBIOTICS

The mass production of antibiotics began during World War II with streptomycin and penicillin. Now most antibiotics are produced by staged fermentations in which strains of microorganisms producing high yields are grown under optimum conditions in nutrient media in fermentation tanks holding several thousand gallons.

The mold is strained out of the fermentation broth and then the antibiotic is removed from the broth by filtration, precipitation and other separation methods. In some cases new antibiotics are laboratory synthesized, while many antibiotics are produced by chemically modifying natural substances; many such derivatives are more effective than the natural substances against infecting organisms or are better absorbed by the body, e.g., some semisynthetic penicillins are effective against bacteria resistant to the parent substance.

Benzylpenicillins and phenoxymethylpenicillins (penicillins G and V, respectively) are produced by fermentation and are the basic precursors of a wide range of semi-synthetic antibiotics, e.g. ampicillin. The amide link may be hydrolysed conventionally but the conditions necessary for its specific hydrolysis, whilst causing no hydrolysis of the intrinsically more labile but pharmacologically essential b-lactam ring, are difficult to attain. Such specific hydrolysis may be simply achieved by use of penicillin amidases (also called penicillin acylases).

Different enzyme preparations are generally used for the hydrolysis of the penicillins G and V, pencillin-V-amidase being much more specific than pencillin-G-amidase. Penicillin amidase may be obtained from E. coli and has been immobilised on a number of supports including cyanogen bromide-activated Sephadex G200.

It represents one of the earliest successful processes involving immobilised enzymes and is generally used in batch or semicontinuous STR processes (40,000 Ukgpenicillin G, 35°C, pH 7.8, 2 h) where it may be reused over 100 times. It has also been used in PBRs, where it has an active life of over 100 days, producing about two tonnes of 6-aminopenicillanic acid $kg^{-1}$of immobolised enzyme.

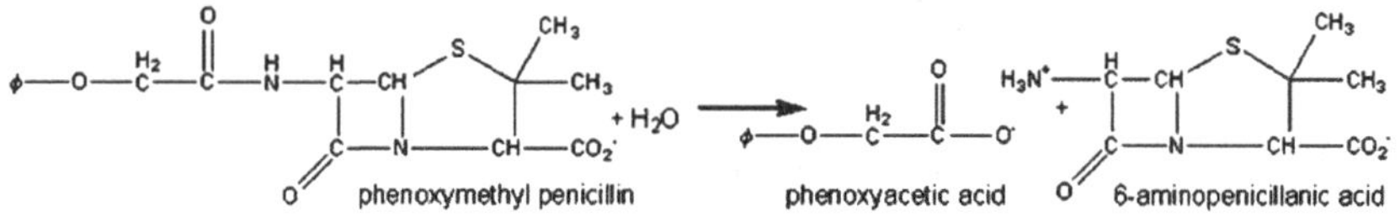

The penicillin-G-amidases may be used 'in reverse' to synthesise penicillin and cephalosporin antibiotics by non -equilibrium kinetically controlled reactions. Ampicillin has been produced by the use of penicillin-G-amidase immobilised by adsorption to DEAE -cellulose in a packed bed column:

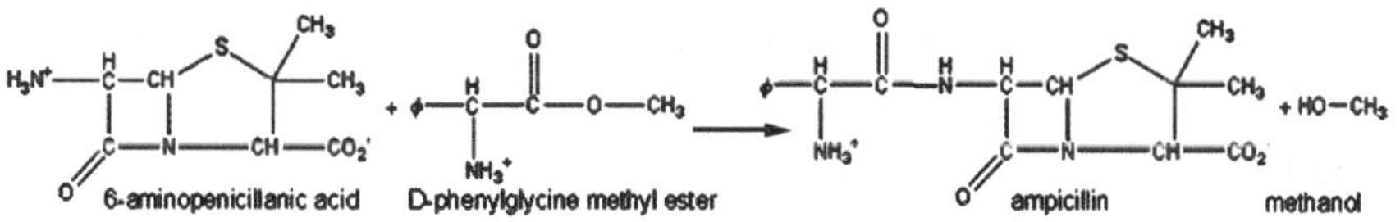

Many other potential and proven antibiotics have been synthesised in this manner, using a variety of synthetic b-lactams and activated carboxylic acids.

PREPARATION OF ACRYLAMIDE

Acrylamide is an important monomer needed for the production of a range of economically useful polymeric materials. It may be produced by the addition of water to acrylonitrile.

$$CH_2 = CHCN + H_2O \rightarrow CH_2{=}CHCONH_2$$

This process may be achieved by the use of a reduced copper catalyst (Cu^+); however, the yield is poor, unwanted polymerisation or conversion to acrylic acid ($CH_2{=}CHCOOH$) may occur at the relatively high temperatures involved (80 –140°C) and the catalyst is difficult to regenerate. These problems may be overcome by the use of immobilised nitrile hydratase (often erroneously called a nitrilase).

The enzyme from Rhodococcus has been used by the Nitto Chemical Industry Co. Ltd, as it contains only very low amidase activity which otherwise would produce unwanted acrylic acid from the acrylamide.

Immobilised nitrile hydratase is simply prepared by entrapping the intact cells in a cross-linked 10% (w/v) polyacrylamide/dimethylaminoethylmethacrylate gel and granulating the product. It is used at 10°C and pH 8.0-8.5 in a semibatchwise process, keeping the substrate acrylonitrile concentration below 3% (w/v).

Using 1% (w/v) immobilised-enzyme concentration (about 50,000 U l^{-1}) the process takes about a day. Product concentrations of up to 20% (w/v) acrylamide have been achieved, containing negligible substrate and less than 0.02% (w/w) acrylic acid. Acrylamide production using this method is about 4000 tonnes per year.

The closely related enzymes cyanidase and cyanide hydratase are used to remove cyanide from industrial waste and in the detoxification of feeds and foodstuffs containing amygdalin.

$$HCN + 2H_2O \rightarrow HCOO^- + NH_4^+$$

$$HCN + H_2O \rightarrow HCONH_2$$

IDENTIFYING USEFUL ANTIBIOTICS

Despite the wide variety of known antibiotics, less than 1% of antimicrobial agents have any medical or commercial value. The most commonly known antibiotic, Penicillin has a highly selective toxicity and therapeutic index (as eukaryotic animal cells do not contain peptidoglycan, they are usually unaffected by it). This is not so for many antibiotics.

Others simply lack advantage over the antibiotics already in use, or have no other practical applications.In order to identify the useful antibiotics, a process of screening is often employed. Using this method, isolates of a large number of microorganisms are cultured and then tested for production of diffusible products which inhibit the growth of test organisms. However, the majority of the resulting antibiotics are already known and must therefore be disregarded.

The remainders must be tested for their selective toxicities and therapeutic activities and the best candidates can be examined and possibly modified.A more modern version of this approach is a rational design programme. This involves screening being directed towards finding new natural products that inhibit specific targets (e.g. a particular step of a metabolic pathway) on microorganisms, rather than tests to show general inhibition of a culture.

Industrial Production Techniques

Antibiotics are produced industrially by a process of fermentation, where the source microorganism is grown in large containers (100,000–150,000 liters or more) containing a liquid growth medium. Oxygen concentration, temperature, pH and nutrient levels must be optimal and are closely monitored and adjusted if necessary. As antibiotics are secondary metabolites, the population size must be controlled very carefully to ensure that maximum yield is obtained before the cells die. Once the process is complete, the antibiotic must be extracted and purified to a crystalline product. This is simpler to achieve if the antibiotic is soluble in organic solvent. Otherwise it must first be removed by ion exchange, adsorption or chemical precipitation.

Strains used for Production

Microorganisms used in fermentation are rarely identical to the wild type. This is because species are often genetically modified to yield the maximum amounts of antibiotics. Mutation is often used and is encouraged by introducing mutagens such as ultraviolet radiation, x-rays or certain chemicals. Selection and further reproduction of the higher yielding strains over many generations can raise yields by 20-fold or more. Another technique used to increase yields is gene amplification, where copies of genes coding for proteins involved in the antibiotic production can be inserted back into a cell, via vectors such as plasmids. This process must be closely linked with retesting of antibiotic production and effectiveness.

TYPES OF ANTIBIOTICS

The great number of diverse antibiotics currently available can be classified in different ways, e.g., by their chemical structure, their microbial origin, or their mode of action. They are also frequently designated by their effective range. Tetracyclines, the most widely used broad-spectrum antibiotics, are effective against both Gram-positive and Gram-negative bacteria, as well as against rickettsias and psittacosis-causing organisms. Ciprofloxacin (Cipro) is another broad-spectrum antibiotic, effective in the treatment of mild infections of the urinary tract and sinuses.

The medium-spectrum antibiotics bacitracin, the erythromycins, penicillin and the cephalosporins are effective primarily against Gram-positive bacteria, although the streptomycin group is effective against some Gram-negative and Gram-positive bacteria. Polymixins are narrow-spectrum antibiotics effective against only a few species of bacteria.

Administration and Side Effects

Antibiotics are either injected, given orally, or applied to the skin in ointment form. Many, while potent anti-infective agents, also cause toxic side effects. Some, like penicillin, are highly allergenic and can cause skin rashes, shock and other manifestations of allergic sensitivity. Others, such as the tetracyclines, cause major changes in the intestinal bacterial population and can result in superinfection by fungi and other microorganisms. Chloramphenicol, which is now restricted in use, produces severe blood diseases and use of streptomycin can result in ear and kidney damage. Many antibiotics are less effective than formerly because antibiotic-resistant strains of microorganisms have emerged.

Nonmedical Use

Antibiotics have found wide nonmedical use. Some are used in animal husbandry, along with vitamin B_{12}, to enhance the weight gain of livestock. However, some authorities believe the addition of antibiotics to animal feeds is dangerous because continuous low exposure to the antibiotic can sensitize humans to the drug and make them unable to take the substance later for the treatment of infection.

In addition low levels of antibiotics in animal feed encourage the emergence of antibiotic-resistant strains of microorganisms. Drug resistance has been shown to be carried by a genetic particle transmissible from one strain of microorganism to another and the presence of low levels of antibiotics can actually cause an increase in the number of such particles in the bacterial population and increase the probability that such particles will be transferred to pathogenic, or disease-causing, strains. Antibiotics have also been used to treat plant diseases such as bacteria-caused infections in tomatoes, potatoes and fruit trees. The substances are also used in experimental research.

INTRODUCTION TO ENVIORNMENTAL BIOTECHNOLOGY

The Chambers Science and Technology Dictionary defines biotechnology as 'the use of organisms or their components in industrial or commercial processes, which can be aided by the techniques of genetic manipulation in developing e.g. novel plants for agriculture or industry.' Despite the inclusiveness of this definition, the biotechnology sector is still often seen as largely medical or pharmaceutical in nature, particularly amongst the general public. While to some extent the huge research budgets of the drug companies and the widespread familiarity of their products makes this understandable, it does distort the full picture and somewhat unfairly so. However, while therapeutic instruments form, in many respects, the 'acceptable' face of biotechnology, elsewhere the science is all too frequently linked with unnatural interference.

While the agricultural, industrial and environmental applications of biotechnology are potentially very great, the shadow of Frankenstein has often been cast across them. Genetic engineering may be relatively commonplace in pharmaceutical thinking and yet in other spheres, like agriculture for example, society can so readily and thoroughly demonise it. The history of human achievement has always been episodic. For a while, one particular field of endeavour seems to hold sway as the preserve of genius and development, before the focus shifts and development forges ahead in dizzy exponential rush in an entirely new direction.

So it was with art in the renaissance, music in the 18th century, engineering in the 19th and physics in the 20th. Now it is the age of the biological, possibly best viewed almost as a rebirth, after the great heyday of the Victorian naturalists, who provided so much input into the developing science.

It is then, perhaps, no surprise that the European Federation of Biotechnology begins its 'Brief History' of the science in the year 1859, with the publication of On the Origin of Species by Means of Natural Selection by

Charles Darwin. Though his famous voyage aboard HMS Beagle, which led directly to the formulation of his (then) revolutionary ideas, took place when he was a young man, he had delayed making them known until 1858, when he made a joint presentation before the Linnaean Society with Alfred Russell Wallace, who had, himself, independently come to very similar conclusions.

Their contribution was to view evolution as the driving force of life, with successive selective pressures over time endowing living beings with optimised characteristics for survival. Neo-Darwinian thought sees the interplay of mutation and natural selection as fundamental. The irony is that Darwin himself rejected mutation as too deleterious to be of value, seeing such organisms, in the language of the times, as 'sports' – oddities of no species benefit. Indeed, there is considerable evidence to suggest that he seems to have espoused a more Lamarckist view of biological progression, in which physical changes in an organism's lifetime were thought to shape future generations.

Darwin died in 1882. Ninety-nine years after his death, the first patent for a genetically modified organism was granted to Ananda Chakrabarty of the US General Electric, relating to a strain of Pseudomonas aeruginosa engineered to express the genes for certain enzymes in order to metabolise crude oil. Twenty years later still, in the year that saw the first working draft of the human genome sequence published and the announcement of the full genetic blueprint of the fruit fly, Drosophila melanogaster, that archetype of eukaryotic genetics research, biotechnology has become a major growth industry with increasing numbers of companies listed on the world's stock exchanges.

Thus, at the other end of the biotech timeline, a century and a half on from Origin of Species, the principles it first set out remain of direct relevance for what has been termed the 'chemical evolution' of biologically active substances and are commonly used in laboratories for in vitro production of desired qualities in biomolecules.

The Role of Environmental Biotechnology

While pharmaceutical biotechnology represents the glamorous end of the market, environmental applications are decidedly more in the Cinderella mould. The reasons for this are fairly obvious. The prospect of a cure for the many diseases and conditions currently promised by gene therapy and other biotech-oriented medical miracles can potentially touch us all. Our lives may, quite literally, be changed.

Environmental biotechnology, by contrast, deals with far less apparently dramatic topics and, though their importance, albeit different, may be every bit as great, their direct relevance is far less readily appreciated by the bulk of the population. Cleaning up contamination and dealing rationally with wastes is, of course, in everybody's best interests, but for most people, this is simply addressing a problem which they would rather had not existed in the first place.

Even for industry, though the benefits may be noticeable on the balance sheet, the likes of effluent treatment or pollution control are more of an inevitable obligation than a primary goal in themselves. In general, such activities are typically funded on a distinctly limited budget and have traditionally been viewed as a necessary inconvenience. This is in no way intended to be disparaging to industry; it simply represents commercial reality. In many respects, there is a logical fit between this thinking and the aims of environmental biotechnology. For all the media circus surrounding the grand questions of our age, it is easy to forget that not all forms of biotechnology involve xenotransplantation, genetic modifica-tion, the use of stem cells or cloning.

Some of the potentially most beneficial uses of biological engineering and which may touch the lives of the majority of people, however indirectly, involve much simpler approaches. Less radical and showy, certainly, but powerful tools, just the same. Environmental biotechnology is fundamentally rooted in waste, in its various guises, typically being concerned with the remediation of contamination caused by previous use, the impact reduction of current activity or the control of pollution.

Thus, the principal aims of this field are the manufacture of products in environmentally harmonious ways, which allow for the minimisation of harmful solids, liquids or gaseous outputs or the clean-up of the residual effects of earlier human occupation. The means by which this may be achieved are essentially two-fold.

Environmental biotechnologists may enhance or optimise conditions for existing biological systems to make their activities happen faster or more efficiently, or they resort to some form of alteration to bring about the desired outcome. The variety of organisms which may play a part in environmental applications of biotechnology is huge, ranging from microbes through to trees and all are utilised on one of the same three fundamental bases – accept, acclimatise or alter. For the vast majority of cases, it is the former approach, accepting and making use of existing species in their natural, unmodified form, which predominates.

The Scope for Use

There are three key points for environmental biotechno-logy interventions, namely in the manufacturing process, waste management or pollution control. Accordingly, the range of businesses to which environmental biotechnology has potential relevance is almost limitless. One area where this is most apparent is with regard to waste.

All commercial operations generate waste of one form or another and for many, a proportion of what is produced is biodegradable. With disposal costs rising steadily across the world, dealing with refuse constitutes an increasingly high contribution to overheads. Thus, there is a clear incentive for all businesses to identify potentially cost-cutting approaches to waste and employ them where possible.

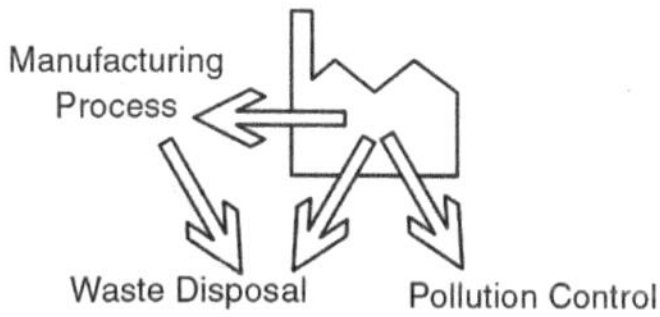

Fig. The Three Intervention Points

Changes in legislation throughout Europe, the US and elsewhere, have combined to drive these issues higher up the political agenda and biological methods of waste treatment have gained far greater acceptance as a result. For those industries with particularly high biowaste production, the various available treatment biotechnologies can offer considerable savings. Manufacturing industries can benefit from the applications of whole organisms or isolated biocomponents.

Compared with conventional chemical processes, microbes and enzymes typically function at lower temperatures and pressures. The lower energy demands this makes leads to reduced costs, but also has clear benefits in terms of both the environment and workplace safety. Additionally, biotechno-logy can be of further commercial significance by converting low-cost organic feedstocks into high value products or, since enzymatic reactions are more highly specific than their chemical counterparts, by deriving final substances of high relative purity.

Almost inevitably, manufacturing companies produce wastewaters or effluents, many of which contain biodegrada-ble contaminants, in varying degrees. Though traditional permitted discharges to sewer or watercourses may be adequate for some, other industries, particularly those with recalcitrant or highly concentrated effluents, have found significant benefits to be gained from using biological treatment methods themselves on site. Though careful monitoring and process control are essential, biotechnology stands as a particularly cost-effective means of reducing the pollution potential of wastewater, leading to enhanced public relations, compliance with environmental legislation and quantifiable cost-savings to the business.

Those involved in processing organic matter, for example, or with drying, printing, painting or coating processes, may give rise to the release of volatile organic compounds (VOCs) or odours, both of which represent environmental nuisances, though the former is more damaging than the latter. For many, it is not possible to avoid producing these emissions altogether, which leaves treating them to remove the offending contaminants the only practical solution. Especially for relatively low concentrations of readily water-soluble VOCs or odorous chemicals, biological technologies can offer an economic and effective alternative to conventional methods. The use of biological cleaning agents is another area of potential benefit, especially where there is a need to remove oils and fats from process equipment, work surfaces or drains.

Aside from typically reducing energy costs, this may also obviate the need for toxic or dangerous chemical agents. The pharmaceutical and brewing industries, for example, both have a long history of employing enzyme-based cleaners to remove organic residues from their process equipment. In addition, the development of effective biosensors, powerful tools which rely on biochemical reactions to detect specific substances, has brought benefits to a wide range of sectors, including the manufacturing, engineering, chemical, water, food and beverage industries.

With their ability to detect even small amounts of their particular target chemicals, quickly, easily and accurately, they have been enthusiastically adopted for a variety of process monitoring applications, particularly in respect of pollution assessment and control.

Contaminated land is a growing concern for the construction industry, as it seeks to balance the need for more houses and offices with wider social and environmental goals. The reuse of former industrial sites, many of which occupy prime locations, may typically have associated planning conditions attached which demand that the land be cleaned up as part of the development process.

With urban regeneration and the reclamation of 'brown-field' sites increasingly favoured in many countries over the use of virgin land, remediation has come to play a significant role and the industry has an ongoing interest in identifying cost-effective methods of achieving it. Historically, much of this has involved simply digging up the contaminated soil and removing it to landfill elsewhere.

Bioremediation technologies provide a competitive and sustainable alternative and in many cases, the lower disturbance allows the overall scheme to make faster progress. As the previous brief examples show, the range of those which may benefit from the application of biotechnology is lengthy and includes the chemical, pharmaceutical, water, waste management and leisure industries, as well as manufacturing, the military, energy generation, agriculture and horticulture. Clearly, then, this may have relevance to the viability of these ventures and biotechno-logy is an essentially commercial activity. Environmental biotechnology must compete in a world governed by the best practicable environmental option (BPEO) and the best available techniques not entailing excessive cost (Batneec).

Consequently, the economic aspect will always have a large influence on the uptake of all initiatives in environmental biotechnology and, most particularly, in the selection of methods to be used in any given situation. It is impossible to divorce this context from the decisionmaking process. By the same token, the sector itself has its own implications for the wider economy.

THE MARKET FOR ENVIRONMENTAL BIOTECHNOLOGY

The UK's Department of Trade and Industry estimated that 15–20% of the global environmental market in 2001 was biotech-based, which amounted to about 250–300 billion US dollars and the industry is projected to grow by as much as ten-fold over the following five years. This expected growth is due to greater acceptance of biotechnology for clean manufacturing applications and energy production, together with increased landfill charges and legislative changes in waste management which also alter the UK financial base favourably with respect to bioremediation.

Biotechnology-based methods are seen as essential to help meet European Union (EU) targets for biowaste diversion from landfill and reductions in pollutants. Across the world the existing regulations on environmental pollution are predicted to be more rigorously enforced, with more stringent compliance standards implemented. All of this is expected to stimulate the sales of biotechnology-based environmental processing methods significantly and, in particular, the global market share is projected to grow faster than the general biotech sector trend, in part due to the anticipated large-scale EU aid for environmental clean-up in the new accession countries of Eastern Europe.

Other sources paint a broadly similar picture. The BioIndustry Association (BIA) survey, Industrial Markets for UK Biotechnology – Trends and Issues, published in 1999 does not quote any monetary sector values per year, but gives the size of the UK sector as employing 40 000 people in 1998 with an average yearly growth over 1995–98 of 20%. Environmental biotech is reported as representing around 10% of this sector. An Arthur Anderson report of 1997 gives the turnover of the UK biotech sector as 702 million pounds sterling in 1995/96, with a 50% growth over three years. A 1998 Ernst and Young report on the European Life Sciences Sector says that the market for biotechnology products has the potential to reach 100 billion pounds sterling worldwide by 2005.

The Organisation for Economic Cooperation and Development (OECD) estimates that the global market for environmental biotechnology products and services alone will rise to some US$75 billion by the year 2000, accounting for some 15 to 25% of the overall environmental technology market, which has a growth rate estimated at 5.5% per annum. The UK potential market for environmental biotechnology products and services is estimated at between 1.65 and 2.75 billion US dollars and the growth of the sector stands at 25% per annum.

An unsourced quote found on a Korean University website says that the world market size of biotechnology products and services was estimated to be approximately 300 billion US dollars in the year 2000. The benefits are not, however, confined to the balance sheet. The Organisation for Economic Cooperation and Development (OECD 2001) concluded that the industrial use of biotechnology commonly leads to increasingly environmentally harmonious processes and additionally results in lowered operating and/or capital costs. For years, industry has appeared locked into a seemingly unbreakable cycle of growth achieved at the cost of environmental damage.

The OECD investigation provides what is probably the first hard evidence to support the reality of biotechnology's long-heralded promise of alternative production methods, which are ecologically sound and economically efficient. A variety of industrial sectors including pharmaceuticals, chemicals, textiles, food and energy were examined, with a particular emphasis on biomass renewable resources, enzymes and bio-catalysis. While such approaches may have to be used in tandem with other processes for maximum effectiveness, it seems that their use invariably leads to reduction in operating or capital costs, or both.

Moreover, the research also concludes that it is clearly in the interests of governments of the developed and developing worlds alike to promote the use of biotechnology for the substantial reductions in resource and energy consumption, emissions, pollution and waste production it offers. The potential contribution to be made by the appropriate use of biotechnology to environmental and economic sustainability would seem to be clear.The upshot of this is that few biotech companies in the environmental sector perceive problems for their own business development models, principally as a result of the wide range of businesses for which their services are applicable, the relatively low market penetration to date and the large potential for growth.

Competition within the sector is not seen as a major issue either, since the field is still largely open and unsaturated. Moreover, there has been a discernible tendency in recent years towards niche specificity, with companies operating in more specialised subarenas within the environmental biotechnology umbrella. Given the number and diversity of such possible slots, coupled with the fact that new opportunities and the technologies to capitalise on them, are developing apace, this trend seems likely to continue.

It is not without some irony that companies basing their commercial activities on biological organisms should themselves come to behave in such a Darwinian fashion. However, the picture is not entirely rosy. Typically the sector comprises a number of relatively small, specialist companies and the market is, as a consequence, inevitably fragmented. Often the complexities of individual projects make the application of 'standard' off-theshelf approaches very difficult, the upshot being that much of what is done must be significantly customised. While this, of course, is a strength and of great potential environmental benefit, it also has hard commercial implications which must be taken into account.

A sizeable proportion of companies active in this sphere, have no products or services which might reasonably be termed suitable for generalised use,

though they may have enough expertise, experience or sufficiently perfected techniques to deal with a large number of possible scenarios. The fact remains that one of the major barriers to the wider uptake of biological approaches is the high perceived cost of these applications.

Part of the reason for this lies in historical experience. For many years, the solutions to all environmental problems were seen as expensive and for many, particularly those unfamiliar with the multiplicity of varied technologies available, this has remained the prevalent view. Generally, there is often a lack of financial resource allocation available for this kind of work and biotech providers have sometimes come under pressure to reduce the prices for their services as a result.

Greater awareness of the benefits of biotechnology, both as a means to boost existing markets and for the opening up of new ones, is an important area to be addressed. Many providers, particularly in the UK, have cited a lack of marketing expertise as one of the principal barriers to their exploitation of novel opportunities.

In addition, a lack of technical understanding of biotech approaches amongst target industries and, in some cases, downright scepticism regarding their efficacy, can also prove problematic. Good education, in the widest sense, of customers and potential users of biological solutions will be one major factor in any future upswing in the acceptance and utilisation of these technologies.

Modalities and Local Influences

Another of the key factors affecting the practical uptake of environmental biotechnology is the effect of local circumstances. Contextual sensitivity is almost certainly the single most important factor in technology selection and represents a major influence on the likely penetration of biotech processes into the marketplace.

Neither the nature of the biological system, nor of the application method itself, play anything like so relevant a role. This may seem somewhat unexpected at first sight, but the reasons for it are obvious on further inspection. While the character of both the specific organisms and the engineering remain essentially the same irrespective of location, external modalities of economics, legislation and custom vary on exactly this basis. Accordingly, what may make abundant sense as a biotech intervention in one region or country, may be totally unsuited to use in another. In as much as it is impossible to discount the wider global economic aspects, disassociating political, fiscal and social conditions equally cannot be done, as the following example illustrates.

In 1994, the expense of bioremediating contaminated soil in the United Kingdom greatly exceeded the cost of removing it to landfill. Six years later, with successive changes of legislation and the imposition of a landfill tax, the situation has almost completely reversed. In those other countries where landfill

has always been an expensive option, remediation has been embraced far more readily.

While environmental biotechnology must, inevitably, be viewed as contextually dependent, as the previous example shows, contexts can change. In the final analysis, it is often fiscal instruments, rather than the technologies, which provide the driving force and sometimes seemingly minor modifica-tions in apparently unrelated sectors can have major ramifications for the application of biotechnology.

The legal framework is another aspect of undeniable importance in this respect. Increasingly tough environmental law makes a significant contribution to the sector and changes in regulatory legislation are often enormously influential in boosting existing markets or creating new ones.

When legislation and economic pressure combine, as, for example, they have begun to do in the European Landfill Directive, the impetus towards a fundamental paradigm shift becomes overwhelming and the implications for relevant biological applications can be immense. There is a natural tendency to delineate, seeking to characterise technologies into particular categories or divisions. However, the essence of environmental biotechnology is such that there are many more similarities than differences.

Though it is, of course, often helpful to view individual technology uses as distinct, particularly when considering treatment options for a given environmental problem, there are inevitably recurrent themes which feature throughout the whole topic. Moreover, this is a truly applied science. While the importance of the laboratory bench cannot be denied, the controlled world of research translates imperfectly into the harsh realities of commercial implementation. Thus, there can often be a dichotomy between theory and application and it is precisely this fertile ground which is explored in the present work.

In addition, the principal underlying approach of specifically environmental biotechnology, as distinct from other kinds, is the reliance on existing natural cycles, often directly and in an entirely unmodified form. Thus, this science stands on a foundation of fundamental biology and biochemistry. To understand the application, the biotechnologist must simply examine the essential elements of life, living systems and ecological circulation sequences. However engineered the approach, this fact remains true. In many respects, environmental biotechnology stands as the purest example of the newly emergent bioindustry, since it is the least refined, at least in terms of the basis of its action.

In essence, all of its applications simply encourage the natural propensity of the organisms involved, while seeking to enhance or accelerate their action. Hence, optimisation, rather than modification, is the typical route by which the particular desired end result, whatever it may be, is achieved and, consequently, a number of issues feature as common threads of individual technologies.

Integrated Approach

Integration is an important aspect for environmental biotechnology. One theme that will be developed throughout this book is the potential for different biological approaches to be combined within treatment trains, thereby producing an overall effect which would be impossible for any single technology alone to achieve. However, the wider goal of integration is not, of necessity, confined solely to the specific methods used. It applies equally to the underpinning knowledge that enables them to function in the first place and an understanding of this is central to the rationale behind this book. In some spheres, traditional biology has become rather unfashionable and the emphasis has shifted to more exciting sounding aspects of life science.

While the new-found concentration on 'Ecological processes', or whatever, sounds distinctly more 'Environ-mental', in many ways and somewhat paradoxically, it sometimes serves the needs of environmental biotechnology rather less well. The fundamentals of living systems are the stuff of this branch of science and, complex though the whole picture may be, at its simplest the environmental biotechnologist is principally concerned with a relatively small number of basic cycles. In this respect, a good working knowledge of biological processes like respiration, fermentation and photosynthesis, a grasp of the major cycles by which carbon, nitrogen and water are recycled and an appreciation of the flow of energy through the biosphere must be viewed as prerequisites.

The intent here has been neither to insult the readership by parading what is already well known, nor gloss over aspects which, if left unexplained, might only serve to confuse. However, this is expressly not designed to be a substitute for much more specific texts on these subjects, nor an entire alternative to a cohesive course on biology or biochemistry. The intention is to introduce and explain the necessary aspects and elements of various metabolic pathways, reactions and abilities as required to advance the reader's understanding of this particular branch of biotechnology.

A large part of the reasons for approaching the subject in this way is the fact that there really is no such thing as a 'typical' environmental biotechnologist. Practitioners come into the profession from a wide variety of disciplines and by many different routes. Thus, amongst their ranks are agronomists, biochemists, biologists, botanists, enzymologists, geneticists, microbiologists, molecular biologists, process engineers and protein technologists, all of whom bring their own particular skills, knowledge base and experiences.

The applied nature of environmental biotechnology is obvious. While the science underlying the processes themselves may be as pure as any other, what distinguishes this branch of biological technology are the distinctly real-life purposes to which it is put. Hence, part of the intended function of this book is to attempt to elucidate the former in order to establish the basis of the latter.

At the same time, as any applied scientist will confirm, what happens in the field under operational conditions represents a distinct compromise between the theoretical and the practically achievable. At times, anything more than an approximation to the expected results may be counted as something of a triumph of environmental engineering.

Closing Remarks

The celebrated astronomer and biologist, Sir Fred Hoyle, said that the solutions to major unresolved problems should be sought by the exploration of radical hypotheses, while simultaneously adhering to well-tried and tested scientific tools and methods. This approach is particularly valid for environmental biotechnology. With new developments in treatment technologies appearing all the time, the list of what can be processed or remediated by biological means is ever changing. By the same token, the applications for which biotechnological solutions are sought are also subject to alteration. For the biotech sector to keep abreast of these new demands it may be necessary to examine some truly 'radical hypotheses' and possibly make use of organisms or their derivatives in ways previously unimagined. This is the basis of innovation; the inventiveness of an industry is often a good measure of its adaptability and commercial robustness.

Bibliography

H. S. Gangwar and Prabhakar Gupta : *A Textbook Of Engineering Mathematics II*, New Age International, Delhi, 2010.

I. J. Nagrath and Madan Gopal : *A Textbook Of Control Systems Engineering*, New Age International, Delhi, 2010.

J.P. Gupta : *A T B of Engineering Mech. and Basic Civil Engineering*, S. Chand Publishing, Delhi, 2011.

John F. James : *Applications in Physics and Engineering*, Cambridge University Press, Delhi, 1998.

K. Venugopal and V. Prabhu Raja : *A Textbook Of Engineering Graphics*, New Age Publications, Delhi, 2009.

N.S. Srinivasan and G. Balasubramanian : *Brain Re-Engineering*, Response Books, Delhi, 2003.

P. C. Sharma : *A Text Book of Production Engineering*, S. Chand Publisher, Delhi, 2010.

R. K. Dhawan : *A Textbook of Engineering Drawing*, S. Chand Publisher, Delhi, 2007.

S. S. Bhavikatti : *A Textbook of Engineering Mechanics*, New Age Publications, Delhi, 2010.

S. S. Dara : *A Textbook of Engineering Chemistry*, S. Chand Publisher, Delhi, 2010.

S.N. Sarbadhikari : *A Short Introduction to Biomedical Engineering*, Universities Press, 2006.

Satyendra Kumar and Nilesh Biwalkar : *Conservation Engineering*, Biotech Books, Delhi, 2014.

Shalin Kumar and Jyoti Rai : *A Textbook of Engineering Chemistry*, Vayu Education of India, Delhi, 2010.

Trymbaka Murthy : *A Textbook of Elements of Mechanical Engineering*, I.K. International Publication, Delhi, 2006.

A. Rashid : *Introduction to Genetic Engineering of Crop Plants*, I.K. International Publication, Delhi, 2009.

A.R. Dabholkar : *General Plant Breeding*, Concept Publication, Delhi, 2006.

Anil Kumar Srivastava : *Genetic Engineering and Biotechnology*, Swastik Publication, Delhi, 2011.

Bipin Kumar Panda : *Encyclopaedia of Biotechnology and Genetic Engineering*, Anmol Publication, Delhi, 2008.

C.P. Malik : *Genetic Engineering : A New Hope for Crop Production and Improvement*, Aavishkar Publication, Delhi, 2010.

Dinesh Arora : *Biotech's Dictionary of Genetic Engineering*, Biotech Books, Delhi, 2004.

Don Grierson : *Plant Genetic Engineering*, Springer Publication, Delhi, 2007.

E. Thro : *Genetic Engineering : Shaping the Material of Life*, Universities Press, Delhi, 2000.

Grierson, Springer : *Plant Genetic Engineering*, Paperback, Delhi, 1998.

Gyan Deep Singh : *Genetic Engineering of Plants*, Anmol Publication, Delhi, 2008.

J.B. Jain and Sumit Jain : *Biotech's Dictionary of Plant Breeding and Genetics*, Biotech, Delhi, 2005.

Jane K. Setlow : *Genetic Engineering : Principles and Methods*, Springer Publication, Delhi, 2010.

L. Yount : *Genetics and Genetic Engineering*, Universities Press, Delhi, 1999.

Ludwig : *Dictionary of Plant Breeding*, Ivy Publication, Delhi, 2008.

M. Follower : *Genetics and Genetic Engineering*, Rajat Publication, Delhi, 2003.

Mark L. Steinberg and Sharon D. Cosloy : *Dictionary Of Biotechnology And Genetic Engineering*, Jaico Publication, Delhi, 1998.

Nicholl : *Introduction to Genetic Engineering*, Cambridge Univ Press, Delhi, 1998.

Nolta : *Genetic Engineering of Mesenchymal Stem Cells*, Springer Publication, Delhi, 1998.

Index